Animal Welfare in a Pandemic

Animal Welfare in a Pandemic explores the impact of COVID-19 on a wide array of animals, from those in the wild to companion and captive animals. During the height of the pandemic, a range of animals were infected, and many died, but this was hard to predict, even using up-to-date bioinformatics. Lockdowns around the world had, and continue to have, a major effect on animals' welfare, influencing pet ownership and care, as well as impacting on the work of conservation institutes due to the lack of visitors and funding and lack of tourist presence in the wild which impacted on anti-poaching efforts. Some of the vast amount of personal protection equipment (PPE) that was distributed was discarded, creating both dangers and occasional opportunities for wild animals. With the rollout of human vaccines, some countries started to develop animal vaccines, only some of which were deployed. In summary, the pandemic had a wide-ranging influence on animal welfare around the world. This is reviewed to highlight what can be learned to protect and enhance animal welfare in future epidemics/pandemics and contribute to a genuinely One Health approach where the health and welfare of both humans and animals are considered holistically.

This book is authored by members of the University of the West of England, Bristol, who span a range of expertise in Biological Sciences, Social Sciences, Animal Welfare, and Ethics.

CRC One Health One Welfare

Learning from Disease in Pets: A 'One Health' Model for Discovery
Edited by Rebecca A. Krimins

Animals, Health and Society: Health Promotion, Harm Reduction and Equity in a One Health World
Edited by Craig Stephen

One Welfare in Practice: The Role of the Veterinarian
Edited by Tanya Stephens

Climate Change and Animal Health
Edited by Craig Stephen and Colleen Duncan

For more information about this series, please visit https://www.routledge.com/CRC-One-Health-One-Welfare/book-series/CRCOHOW

Animal Welfare in a Pandemic

What Does COVID-19 Tell Us For the Future?

John T. Hancock

Ros C. Rouse

Tim J. Craig

CRC Press

Taylor & Francis Group

Boca Raton London New York

CRC Press is an imprint of the
Taylor & Francis Group, an **informa** business

Designed cover image: Ros Rouse

First edition published 2024
by CRC Press
2385 NW Executive Center Drive, Suite 320, Boca Raton FL 33431

and by CRC Press
4 Park Square, Milton Park, Abingdon, Oxon, OX14 4RN

CRC Press is an imprint of Taylor & Francis Group, LLC

© 2024 John T. Hancock, Ros C Rouse, Tim J. Craig

Library of Congress Cataloging in Publication Data
[Insert LoC Data here when available]

ISBN: 9781032547343 (hbk)
ISBN: 9781032521091 (pbk)
ISBN: 9781003427254 (ebk)

DOI: 10.1201/9781003427254

Typeset in Bembo
by Deanta Global Publishing Services, Chennai, India

*"The lower animals, like man, manifestly feel pleasure and
pain, happiness and misery."*

Charles Darwin

*"The greatness of a nation and its moral progress can be judged
by the way its animals are treated."*

*"I hold that the more helpless a creature the more entitled it is
to protection by man from the cruelty of humankind."*

Attributed to Mahatma Gandhi

Contents

Authors

John T. Hancock

John is Professor of Cell Signalling at the University of West of England, Bristol (UWE), UK. He studied at the University of Bristol where he obtained a BSc (Hons) in Biochemistry (1984) and then a PhD (1987). He stayed at Bristol where he held post-doctoral research positions for six years, subsequently moving to UWE in 1993. He has been at UWE ever since. John has a long-standing interest in how small relatively reactive molecules are able to partake in cell signalling events in cells. This often centres around the chemistry of reduction and oxidation (redox). He has authored several editions of the textbook *Cell Signalling*, published by Oxford University Press, as well as publishing over two hundred academic articles. He holds several editorial positions with journals, most notably being the Editor-in-Chief of the journal *Oxygen*. Recently, John has published the book *Why Elephants Cry: How Observing Unusual Animal Behaviours Can Predict the Weather (and Other Environmental Phenomena)* with CRC Press. He has also written several articles on COVID-19, including about the impact of the pandemic on animals and animal welfare, and it was this that initiated the writing of this book.

Ros C. Rouse

Ros originally studied Psychology at the University of Bristol. She is a science policy expert with extensive Research Council experience. Ros currently concentrates on promoting integrity in research at the University of the West of England and is a member of the University's Ethics and Integrity Committee. She has a strong personal commitment to the welfare of animals and believes that every animal counts. Ros is committed to a vision of One Health in which the health and well-being of animals is considered, as well as the impact of their health on humans. Ros has co-authored several articles on COVID-19 and animals. Ros is also an artist, focussing on wildlife and the natural world.

Tim J. Craig

Tim originally studied Molecular and Cellular Biochemistry at the University of Oxford, graduating in 2000, before studying for a PhD at the University of Liverpool on the molecular mechanisms of neurotransmitter release. His first postdoctoral position was in ion channel physiology at the University of Oxford (2004–2009) followed by a senior post-doctoral position at the University of Bristol (2009–2015), during which he worked on many aspects of molecular neuroscience, including synaptic plasticity and neurotransmitter release. Since 2015, he has held an academic position at the University of the West of England, Bristol, where he is currently an Associate Professor of Neuroscience. In this role he teaches many different courses and provides much of the neuroscience teaching on the Biological and Biomedical Sciences degree courses. His research currently focusses on the role of dietary factors in neuronal health and function.

Preface

The COVID-19 pandemic, which started in 2019, has been a global event with significant consequences for the human population. Millions of people were ill and lost their lives. Others lost their jobs and endured long lockdowns. People suffered financially and mental health declined for many. The impact on many communities around the world should not be underestimated. However, what happened to the animals?

The pandemic had a significant impact on many animal species and communities. Many died directly by being infected, whilst others suffered indirectly through the actions taken by humans to mitigate the effects of the pandemic. Much of this was not reported, with the media focussed on human infections and deaths. This focus saw reactive measures being implemented to protect people, with often little thought given to the consequences for the ecosystems around us. Here, aspects of how the pandemic had an impact on animals will be discussed, and the text will also look to the future to suggest that lessons can be learnt from the action taken because of the COVID-19 pandemic over the last three years.

The book starts with an overview of the COVID-19 pandemic and a brief history of similar viral epidemics, such as SARS and MERS. In the next chapter the question will be asked as to whether molecular biology may be able to predict which animal species would be susceptible to these coronaviruses. A reality check is the focus of Chapter 3, where the discussion centres on the direct impact of the virus on animals. Chapter 4 deals with animal vaccines. Subsequent chapters focus more on the indirect effects of the pandemic, for example, the impact on animal conservation and the danger of abandoned personal protective equipment. The last chapter discusses whether any lessons can be learnt from the actions taken during the 2019 pandemic that can be used in the future, as, no doubt, COVID-19 will not be the last epidemic/pandemic to occur.

This book is not intended to give a systematic coverage of all the material on this topic but rather to overview some of the issues that arose from the pandemic and how this impacted animals. With that in mind, although the majority of the literature is recent, as it is pertinent to recent events, we used no cut-off date for citations. We also drew on a broad coverage that included media outlets from Internet pages and newspapers as well as the peer-reviewed science literature. We hope that we have given any credit due for use of all sources. Any mistakes and misinterpretations are our own.

Finally, we would like to thank all those who have contributed literature on this topic or have discussed this with us as we prepared the manuscript, and we hope you enjoy the text that follows.

John T. Hancock, Ros C. Rouse, and Tim J. Craig

Acknowledgements

No book is written in isolation. We would like to thank all those who have supported us as we put this text together.

We acknowledge the photographs, which were mainly acquired from Shutterstock. We appreciate the massive effort by people who make the bioinformatic tools that we use, such as Clustal Omega, and the database entries at www.rcsb.org. Other images were our own.

We would particularly like to thank Alice Oven, who invited us to propose this book, and her amazing team at CRC Press, and Amelia Bashford who took over the editorial process for this book.

Lastly, we would like to thank our families for their patience with us whilst we worked on this book. As always, John would like to thank Sally-Ann, Thomas, and Annabel. Ros would like to thank Paul. Without their support such texts would never appear.

The coronavirus disease 2019 (COVID-19) pandemic

1.1 Introduction

How would one describe the COVID-19 pandemic? Global – certainly. Disruptive, dangerous, frustrating, controlling, unnecessary? A surprise? Predictable, as some had already said such an event would take place?[1]

However, it is doubtful that a word involving animals would come to mind, with the exception of theories as to the origin of this virus. The pandemic has been a very anthropocentric event. We have been following infection and death rates of humans, and rightfully so. Newspaper and television news almost seemed obsessed with informing people how the pandemic was developing and affecting society, with constant reports of hospitals being overwhelmed, oxygen supplies being limited, and, sadly, corpses mounting up. Of course, this is understandable, but how many reports on the effects on animals were published or broadcast? But animals were affected, and many died.

Coronavirus disease 2019 (COVID-19) is a human disease, i.e., it infected and caused symptoms in humans. However, it is often referred to as a "zoonotic" disease because it was thought to have transferred to the human population via one or more animal species. Having said that, it is probably technically incorrect to say that animals are suffering from COVID-19. The disease is caused by a virus, known as SARS-CoV-2, and therefore when discussing animals, it is probably more correct to say that they are infected by the virus, and this is the position taken throughout this book.

On many occasions during lockdowns, people escaped when they could. Many went walking their dogs – dog ownership seemed to become a trend. It was good to see people standing in groups, perhaps with friends and colleagues, having a normal social life. However, they were invariably socially distancing, keeping the magical two metres apart. But what of the canines with them? They could not read the social distancing rules and would have not understood it even if they could. They milled around, rubbed up against each other, licking various body parts. One can't help wondering if this was wise. Did people with COVID-19, even if asymptomatic, deposit the virus on the surface of their dogs, only to be spread to other dogs in the neighbourhood? On the way home these owners would have patted and stroked their beloved pets. Were they then being exposed to a new source of the virus? Was this ever discussed as an issue? Should it have been?

The news and the vast majority of the research carried out on COVID-19 was focused directly on human health. If animals were discussed it was as the source of the virus back in December 2019. Very little was reported about animals becoming infected or dying from infection. But it was occurring, and it will continue – the pandemic is not over (written at the end of 2022/spring–summer 2023). Here, we have attempted to bring together some of the evidence about animals during the pandemic, and we will end by giving our thoughts on how we might do things better in the future.

Of course, COVID-19 was not the first epidemic/pandemic to be caused by a virus, with some other examples being listed in Table 1.1. Many of these also involved an animal at some stage, either as the source of the virus or as a method of transmission.

1.2 COVID-19 pandemic and the virus that caused it

Coronavirus disease 2019 (COVID-19) was first reported in China in December 2019. However, there were reports in other countries around the same time. For example, in December 2019, samples of sewage in Milan and Turin, Italy, were reported to be positive for the coronavirus, before the first case of COVID-19 was reported from China.[7] Even earlier than this were reports from Brazil, where sampling of sewage showed that the virus was present on 27 November 2019,[8] whilst in Spain there was some evidence of the virus being present as early as March 2019.[9] Wherever, or whenever, it started, in March 2020 it was declared a pandemic. It had gone global at an alarming rate, spreading as people travelled across national borders, particularly by plane. Some researchers have tried to understand the situation using mathematical models,[10] so mitigating action may be taken, or not, if deemed not to be worth doing.

COVID-19 is caused by SARS-CoV-2, which is a coronavirus. This type of virus was first imaged by June Almeida (née Hart). Born in

Table 1.1 Examples of infections that caused an epidemic/pandemic and the animals that were involved as origins or as transmitters of the virus

Disease	Comments	Animal(s) involved
Bubonic plague[3]	Bacterium (*Yersinia pestis*) not virus. Probable cause of "Black Death" 1346–1353, which killed 40%–60% of European population (50 million human deaths)	Fleas from small animals (most notably rats)
"Spanish flu," otherwise known as the Great Influenza epidemic	First described in 1918 50–100 million human deaths[4]	Pigs implicated but debated[5]
Rift Valley Fever	First described in 1931 Other outbreaks in 1977, 1997, and 2000	Contact with body parts of infected animals, transmitted by mosquitos
West Nile Fever	First described in 1937	Transmitted by mosquitos
HIV/AIDS	First documented infections 1959–1960. Worldwide spread from late 1960s (but probable origin in Democratic Republic of the Congo in 1920s). >40 million deaths[6]	Probably from wild ape populations of West Central African forests
Marburg Haemorrhagic Fever	First described in 1967 About 470 deaths	African green monkeys
Lassa Fever	First described in 1969 Annual death rate about 5,000	Rodents
Avian Flu	First described in 1996 Some human deaths	Birds, particularly poultry
Monkeypox	First described in humans in 1970 in the Democratic Republic of the Congo	Described in several mammals, including monkeys and rodents
Ebola virus disease	First appeared in 1976. 50% mortality – sometimes called the "most deadly virus"	Involves several mammals including bats, primates, and antelope
Nipah virus	First identified in 1999	Affects a range of animals

(*Continued*)

Table 1.1 (Continued) Examples of infections that caused an epidemic/pandemic and the animals that were involved as origins or as transmitters of the virus

Disease	Comments	Animal(s) involved
SARS-CoV	First described in 2002 About 813 human deaths	Bats implicated
"Swine flu"	First described in 2009 Estimated as many as 284,000 deaths	Pigs
MERS-CoV	First described in 2012 About 935 deaths	Camels
Chikungunya	First described in 2014, but rare	Transmitted by mosquitos
Zika	First described in 2015	Transmitted by mosquitos
COVID-19	First described in 2019 Caused by SARS-CoV-2 Many millions of deaths	Origin bats, but many other animals may become infected

Data drawn partly from Excler et al.[2]

Glasgow in 1930, she was the daughter of a local bus driver and left school at the age of 16. However, she found a career in electron microscopy. Having spent some time in Canada, she moved to St Thomas's Hospital in London, where in 1964 she captured the first images of a virus that looked like a crown. It was round with spikes sticking out of its surface and the name coronavirus was born – *corona* is Latin for crown. The work stemmed from investigations of a virologist called David Tyrrell, who was working at the Common Cold Unit in Salisbury, Wiltshire, England. He was particularly interested in a virus labelled as B814, which had been obtained from a nasal swab from a boy who had the common cold. Tyrrell gave a sample to Almeida and history was made. This work was first published in the *British Medical Journal* (*BMJ*) in 1965,[11] and then there was a paper in the journal *Nature* in which the term coronavirus was first used[12]. Both Almeida and Tyrrell were amongst the eight authors, who say in their paper:

> In the opinion of the eight virologists these viruses are members of a previously unrecognized group which they suggest should be called the coronaviruses, to recall the characteristic appearance by which these viruses are identified in the electron microscope. (Figures 1.1 and 1.2).

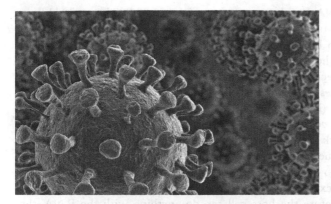

Figure 1.1 An artist's impression of a coronavirus. The coronavirus is so named as it has spike proteins making it appear as a corona or crown. Image from Shutterstock, courtesy of creativeneko. This image has had 3D rendering and has been made black and white.

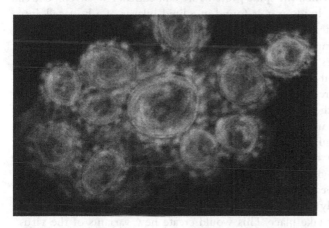

Figure 1.2 A group of SARS-CoV-2 viruses as seen in an electron microscope. Image from Shutterstock, courtesy of Pong Ch. Image has been made black and white.

Transmission Electron Microscope (TEM) images of SARS-CoV-2 have been published, e.g., by Prasad et al.[13]; however, they tend not to be so impressive as the images seen in popular articles and on the television. These are often digitally created, and digitally coloured, so they look good as a backdrop for the news.

The SARS-CoV-2 virus is also classified as an RNA virus. Das et al.[14] state that "Coronaviruses are enveloped, positive-sense, single-stranded RNA virus with 3′ and 5′ cap structure." This means that the virus contains an RNA-based genome, approximately 30,000 bases (30 kb[15]) in length.[16] To be useful for the host cells' molecular machinery this RNA would need to be reverse transcribed into DNA before the host cell could express the genes in the viral genome. Interestingly, this principle is used in molecular biology as a technique to measure the presence of RNA in samples, a technique known as reverse transcriptase polymerase chain reaction (RT-PCR) – "reverse transcriptase" is the enzyme that makes a DNA copy from an RNA molecule. Once inside the cell, the virus can hijack the transcriptional and translational machinery of the cell and make new versions of the virus. These can then be released and spread, either to other cells in the host or to new hosts. Hence, the virus can sustain its existence and propagate the disease.

However, before the host cell can do anything with the RNA, the virus has to enter the host cell. This is where the spike proteins on the virus surface come in. It was these spike proteins that caught Almeida's eye in her electron micrographs. The RNA is encased in a membrane-like envelope into which the spike proteins are embedded. The envelope itself is composed of lipids, just as found in a lipid-bilayer of a host cell, into which the proteins are embedded. Therefore, this lipid was an Achilles' heel for the virus as it could be targeted by materials that destroy lipid-based structures. Hence, detergents were an effective way to "kill" the virus and stop it spreading.[17] However, it is the spike proteins that allow the virus/host recognition. As discussed in Chapter 2, it is the amino acid sequence and the three-dimensional structure of the polypeptides sticking out of the virus and the concomitant proteins sticking out of the host cell that determine whether the virus can infect its target host, or not.

There are two things we need to consider whilst discussing the mechanism of viral "life,"[18] both of which will be revisited below. First, as the virus goes through the cycle of host recognition, cell entry, and then pertinently being copied into new versions, mutations in the RNA sequences may take place. This would create new variants of the virus. Some of these may not be viable, as they can no longer recognise the host or hijack the host's molecular mechanisms, but other variants may be better – as far as the virus is concerned – than the original. As the pandemic continued, hundreds of variants of SARS-CoV-2 were identified, but only a few made the press, as they were the ones of concern, either being more easily spread or causing worse symptoms in the host. Second, the virus, and any variants, can use the same mechanisms to infect cells that were not its original specific target. Therefore, human-targeting viruses may also infect other species. The symptoms may not be the same. Some species may show less severe symptoms, but some species may have the full gamut of symptoms and may, like humans, die from the infection. Into this mix, it has to be remembered that as variants were created, the infection of hosts, and the severity of the disease, may be different across animal species. Just because a variant may be less likely to infect a human does not automatically mean that that same virus variant will not be better at infecting, or having more severe symptoms, in a dog.

It is worth pointing out, as it caused both consternation and much discussion at the time, that it was quite early in the pandemic when it was found that not all people infected with SARS-Cov-2 had symptoms; they were asymptomatic.[19] This caused a problem as it was not obvious who was spreading the disease. People were not testing for infection if they were feeling fine, and then they could go out and spread the virus. Obviously, lockdowns partly solved this problem, but even during lockdowns many people who had essential roles in society were allowed to travel and work, whilst others may not have observed the lockdowns as strictly as they should. Of pertinence here are the animals. If humans could be asymptomatic, animals could too. Therefore, your friendly dog at home could have potentially been infected, but you would not know. Such issues will be discussed in more depth in the subsequent chapters.

It is therefore pertinent to look at the genetic makeup of humans, or any other species being infected. Exploration of the genomes of many people who have been infected with COVID-19 have highlighted specific gene variants that may lead to the disease symptoms being more severe.[20] A multi-group effort published in *Nature* in May 2023[21] indicated 49 such genes, 16 of which had not been found before. Such data will allow better treatments and management of the pandemic, and perhaps similar coronavirus outbreaks in the future. Of course, similar studies in animal species would be useful here too. It was not a trivial undertaking, just for one species (humans), and therefore such studies are unlikely to be carried out in the near future on animals, but if a species was found to have a large infected population, the diversity of the intra-species genetics and how that might lead to either death or asymptomatic disease would be useful to know.

1.3 The origin of the SARS-CoV-2

SARS-Cov-2 almost certainly emanated from a bat species, probably a horseshoe bat (*Rhinolophus*). There are at least 106 species of these bats, and in China several are common, including the greater horseshoe bat (*R. ferrumequinum*), the Chinese rufous (*R. sinicus*), the big-eared (*R. macrotis*), the woolly (*R. luctus*), and the king (*R. rex*[22]). However, the origin of the SARS-CoV-2 virus was possibly the intermediate horseshoe (*R. affinis*). Such viruses are often referred to as being zoonotic, as they are transmitted from an animal species to humans. It is thought that the bats were living in a cave system in southwest China, in Yunnan Province, Mijiang

Figure 1.3 Intermediate horseshoe bat (*Rhinolophus affinis*). Image from
Shutterstock curtesy of Binturong-tonoscarpe, but made black and white.

County, but somehow the virus was transferred to the human population. There has been considerable debate about the true origin of the virus, and the true story may never be unravelled, although an interesting attempt was made by Chan and Ridley[23] (Figure 1.3).

Whilst the first reported cases of COVID-19 were in the city of Wuhan, China, there are two things to note here. The exact origin of the virus has still not been identified, and this has not been the first example of a widespread disease being caused by similar coronaviruses: COVID-19 followed in the footsteps of SARS and MERS.

The first reports suggested that the virus emanated from a "wet market" in Wuhan. These "wet markets" are markets where many kinds of local produce are sold, including fresh food and live animals, which can be slaughtered for food. There are often a range of live animals at such places. Although not an unbiased source, *Animals Equality UK* infiltrated many wet markets and on their website state:

> Turtles, chickens, frogs, ducks, geese, pigeons, fish and more are traded and killed, bringing together species that would never live together in the wild.

Furthermore, sanitary conditions are not excellent. In an on-site assessment of hygiene and biosecurity the following results were obtained when stalls selling birds (and some amphibians) were investigated[24]:

Hygiene rating:

- 4.5% excellent
- 2.3% good

- 50% moderate
- 43.2% poor (needing major improvement)

The authors used a measure of hygiene known as Hygiene and Biosecurity Assessment Tool (HBAT). The results show that huge improvements are required, at least in the wet market studied. One recommendation was that handwashing and toilet facilities were improved, both for stallholders and for customers. Similar outcomes were obtained in a study in Hong Kong,[25] where surfaces were found to be low in hygiene standards and there was a concern that there could be spread of organisms significant for human health, such as *Klebsiella pneumoniae*.

The emergence of human disease from wet markets is not a new phenomenon. In 2004 Robert Webster noted that a wet market was the source of the H5N1 bird-influenza virus, an outbreak that killed six people in Hong Kong.[26] This was not a lone view. In 2006 Woo et al.[27] concluded the summary of their paper with the statement:

> The wet-markets, at closer proximity to humans, with high viral burden or strains of higher transmission efficiency, facilitate transmission of the viruses to humans.

Others point out that wet markets are an important source of food in many countries, but they still recognise that there is a risk of emergence of human disease at these places, including food-borne pathogens.[28]

Xiao and colleagues[29] had carried out a survey of animals that were sold at Wuhan wet market before the pandemic started, i.e., May 2017 to November 2019. They counted 38 species, and, amongst these, 31 were protected. However, they found no evidence of bats being sold, and interestingly no evidence of pangolins, suggesting that the latter was not an intermediate animal between bats and humans. These authors, too, highlight the poor animal welfare seen at wet markets, and they also list a range of potential diseases that may emanate from such places in the future. They say that "wildlife is thought to be the source of at least 70% of all emerging diseases." This is discussed further in section 7.3.

A study using metagenomic[30] sequencing data publicly available on the GISAID database,[31] set out to analyse which animals may have been available at the Wuhan wet market and, hence, could have been a source of SARS-CoV-2, or at least acting as an intermediate animal. In the highlights of the paper[32] the authors state that:

> This finding corroborates reports of putative intermediate animal hosts for SARS-CoV-2 being sold live in the market in late 2019 and adds to the body of evidence identifying the Huanan market

as the spillover location of SARS-CoV-2 and the epicenter of the COVID-19 pandemic.

The authors go on to state that they could identify:

raccoon dog (*Nyctereutes procyonoides*), Malayan porcupine (*Hystrix brachyura*), Amur hedgehog (*Erinaceus amurensis*), masked palm civet (*Paguma larvata*), and hoary bamboo rat (*Rhizomys pruinosus*) from wildlife stalls positive for SARS-CoV-2.

However, it was raccoon dogs that became more of a focus of this work. Evidence reported in March 2023 suggested that one of the intermediate animals involved in the transfer of SARS-CoV-2 to humans was indeed the raccoon dog.[33] There is good evidence that this species was being sold at the Wuhan market and could have led to humans being infected.[34] Evidence was given that the DNA from the mitochondria of these animals were found in six samples that had been collected from two stalls. This is not direct evidence of viral transfer but shows that the animals were present. It is interesting to note that, experimentally, raccoon dogs can be infected with the SARS-CoV-2 virus and that animal-to-animal transmission can take place, with the dogs showing no symptoms, although viral replication can take place in the noses of these animals.[35] This is not proof of where the SARS-CoV-2 comes from, but does open up possibilities and helps future investigations.

Andersen et al.[36] rule out the possibility that the virus originated in a laboratory. They say that receptor-binding domain (RBD) of the viral sequence is optimised in a way that would not have been predicted, suggesting that it was not created by researchers. They suggest that mutations, and therefore natural selection, in the virus would have occurred either in an animal and unfortunately resulted in a virus that was able to infect humans or in a human who was infected and the mutations and optimisation of the virus for further transmission occurred whilst it was circulating in the human population. Either way, they suggest that this was a natural phenomenon and not laboratory created. On the other hand, on 28 February 2023, Director of the FBI Christopher Wray was quoted as saying that he "acknowledged that the bureau believes the Covid-19 pandemic was likely the result of a lab accident in Wuhan, China."[37] This view is shared by the US Department of Energy but, interestingly, not by several other US government departments, highlighting the considerable uncertainty as to the origin of this virus. In June 2023, an article in *The Times* gave a comprehensive overview of evidence of the origin of the virus. In the highlights it is stated:

Fresh evidence drawn from confidential files reveals Chinese scientists spliced together deadly pathogens shortly before the pandemic, the Sunday Times Insight team report.[38]

This gives the definite impression that the virus emanated from a Wuhan laboratory, but this cannot be proved, and it is certain that the argument of the exact origin of SARS-CoV-2 will continue for sometime to come, and it will probably never be satisfactorily resolved.

The WHO at the time (early March 2023) called for all countries that had information about the origins of the virus to share their information. The WHO director general, Tedros Adhanom Ghebreyesus, was quoted as saying that there was a need to

advance our understanding of how this pandemic started so we can prevent, prepare for and respond to future epidemics and pandemics.[39]

It may never be known where SARS-COV-2 originated or how it jumped from an animal to humans. China temporarily closed wet markets, which was thought to be a good thing. However, despite the worry about wet markets, it appears that they are reopening as the pandemic fades.[40] Lin et al.[41] identified six points of concern with regard to wet markets. These were:

(1) presence of high disease–risk animal taxa;
(2) presence of live animals;
(3) hygiene conditions;
(4) market size;
(5) animal density and interspecies mixing;
(6) the length and breadth of animal supply chains.

And then they call for a better classification of these markets so that human health can be prioritised. As pointed out by Nadimpalli et al.,[42] WHO has developed guidance for the reopening of wet markets,[43] and hopefully these guidelines will be adhered to. Nadimpalli et al. conclude by saying:

Using monitoring data to target wet markets for hygiene and sanitation infrastructure upgrades, while protecting these marketplaces as vibrant, affordable, community spaces should be the global public health community's next major focus.

Others call for a ban on the sale of exotic animals at wet markets.[44] This would not only be good for the prevention of potential future pandemics,

Figure 1.4 A painting of a pangolin by Ros Rouse© (made black and white). The originalcanbefoundathttps://www.tigerloveart.co.uk/art/pangolin-bathing-by-ros -rouse-copy-copy/ (Accessed 12/04/23).

but would also be good for populations of such animals, which are often endangered (Figure 1.4).

It is clear from the debate in both the media and the scientific litera- ture that it is unlikely that the origin of SARS-CoV-2 will be uncovered, but that it is likely that a wet market in Wuhan was involved. Opening the debate about the future of such markets can only be good. Hopefully, the types of animals and animal products sold will be monitored, or curtailed in some cases, and hygiene will be improved. It is unlikely that such wet markets will cease to exist, as they are an integral part of many communities, but better monitoring and regulation will go a long way to preventing one of these places being the source of the next pandemic.

There are two origins of COVID-19: Either the virus escaped from a laboratory in Wuhan or it was natural. In the latter case it would have "jumped" from a bat to humans, but almost certainly via one or more intermediate animal species. Originally pangolins were thought to be such an intermediate, but as discussed above, pangolins did not seem to be present in the Wuhan wet market, making them a doubtful intermediate in this process. Others have pointed to other animals, e.g., racoon dogs and foxes.[45] Other species implicated include minks or ferrets, turtles, snakes, yaks, dogs, cats, and pigs. Snakes and birds were ruled out, as were civets and other carnivorous mammals. Pangolin were thoughts to be unlikely, as were livestock, but it was thought that rodents were worth considering.[46] A good summary of the evidence for and against any of these animals being involved was tabulated by Zhao et al.[47] It may never be known for sure.

Of particular pertinence here is that the genome of the virus is not static, i.e., it can change. Over the months since the SARS-CoV-2 outbreak was first reported there have been numerous variants isolated. Many are of little significant importance, as they differ very little from the original. Other variants are of more consequence, as they may cause different symptoms or be transmitted through the population more easily. As of April 2022, the CDC in the United States lists variants of SARS-CoV-2 under four categories, as shown in Table 1.2.

It is reassuring that there are no variants of high consequence listed in Table 1.2. Of course this, as usual, is a very anthropocentric view, as is typical in these listings. Such viral variants may have different consequences for animals, perhaps allowing them to become infected, whereas before they were not able to contract the disease. New variants may cause symptoms where before the virus was seen as asymptomatic, or perhaps new variants are able to transmit the disease more easily. More worrying is whether the production of any new variants in animals can re-emerge

Table 1.2 The most well-known variants of SARS-CoV-2. Data obtained from the CDC web pages[48]

Classification	Variant
Variant being Monitored (VBM)	Alpha (B.1.1.7 and Q lineages)
	Beta (B.1.351 and descendent lineages)
	Gamma (P.1 and descendent lineages)
	Delta (B.1.617.2 and AY lineages)
	Epsilon (B.1.427 and B.1.429)
	Eta (B.1.525)
	Iota (B.1.526)
	Kappa (B.1.617.1)
	Mu (B.1.621, B.1.621.1)
	Zeta (P.2)
	1.617.3
Variant of Interest (VOI)	None listed
Variant of Concern (VOC)	Omicron (B.1.1.529, BA.1, BA.1.1, BA.2, BA.3, BA.4 and BA.5 lineages)
Variant of High Consequence (VOHC)	None listed

back into the human population as a more significant virus, perhaps transmitting between people more easily, or causing worse symptoms, including death. The generation of new viral variants has significant ramifications for the efficacy of any vaccines produced too, as discussed in Chapter 4.

1.3.1 SARS: origins and impact

Before the COVID-19 pandemic probably the most well-known coronavirus epidemic was SARS. This first appeared in November 2002 in the Guangdong Province of China. Within a year 8,076 people were known to have become infected and 778 died.[49] It was not completely isolated but spread across 28 regions, as well as spreading to four other countries, including Canada. In Hong Kong the virus was prevalent from March to May 2003, with 1,718 cases reported. There was an estimated death rate of approximately 15%: 253 people died.[50] The World Health Organisation (WHO) says that the fatality rate was only about 3%, and that most people affected were healthy adults between the ages of 25 and 70. They also said that the virus was airborne and was spread by aerosol, but that people could pick up the virus from surfaces too.[51] In 2009 a study about mainland China gave the cases and deaths as 5,327 and 343, respectively, with a fatality rate of 6.4%.[52] As mentioned, in total it was thought that over 8,000 people were infected, with at least 700 deaths[53] (although a higher number is used above).

SARS was a respiratory disease, and this is reflected in the symptoms seen. The incubation period could be up to ten days, but it could be as short as two days, with the first symptom usually being fever. This may have been accompanied with headaches and muscle pain, and patients often had a dry, non-productive cough. Some people had shortness of breath, and in some people white blood cell and platelet counts were reduced. In severe cases patients needed hospitalisation and, as mentioned, some people died.

The virus involved was a coronavirus that was named severe acute respiratory syndrome-associated coronavirus (SARS-CoV), hence, the name of the COVID-19 virus as SARS-CoV-2. Like the major theories surrounding COVID-19, the SARS virus probably originated in a bat.[54] Also, in a similar manner to COVID-19, it is thought that the intermediate animals were from markets, and species suggested include Himalayan palm civets (*Paguma larvata*), Chinese ferret badgers (*Melogale moschata*), and raccoon dogs (*Nyctereutes procyonoides*).[55]

SARS occurred over 15 years before COVID-19, so were any lessons learnt? Feng et al.[56] suggested that China needed better reporting of possible disease outbreaks and better collaboration between medical departments, as well as more training in disease control. Hung[57] asked in 2003 what lessons could be learnt from SARS. It was clear that communication about the disease was lacking, as was the infrastructure needed to contain it. Medical workers were put at risk and there was a lack of awareness of the impact that may ensue. It was concluded that "There is a need to strengthen the exchange of epidemiological information on infectious diseases, especially the emergence of new infections, between the health authorities in Mainland China and Hong Kong." In light of COVID-19, one could suggest that this should include better communications with the rest of the world too.

SARS was the first major outbreak of a coronavirus and caused numerous deaths. It may have been thought that this might have been a wake-up call for the rest of the world, but it was not the only epidemic that should have been a warning. In 2012 another coronavirus also caused a problem. This time it was referred to as MERS.

1.3.2 MERS: origins and impact

Middle East respiratory syndrome coronavirus (MERS-CoV) caused an epidemic in 2012. The epidemic started in Saudi Arabia, and the countries in the immediate region known to be affected, according to the Centers for Disease Control and Prevention (CDC), are listed as "Bahrain, Iran, Jordan, Kuwait, Lebanon, Oman, Qatar, Saudi Arabia, United Arab Emirates (UAE), and Yemen."[58] Other countries further afield that were also known to have cases of MERS include "Algeria, Austria, China, Egypt, France, Germany, Greece, Italy, Malaysia, Netherlands, Philippines, Republic of Korea, Thailand, Tunisia, Turkey, United Kingdom (UK), and United States of America (USA)." A World Health Organisation (WHO) report says that 27 countries were affected, and that there were 858 known deaths.[59] It was therefore not an isolated disease outbreak and had worldwide consequences, although it was never declared as a pandemic.

For most people the symptoms were fever, a cough, and shortness of breath, but people with an underlying precondition that may weaken their immune system were more likely to have more severe symptoms, and, as mentioned, many people died (Figure 1.5).

The virus, as was the case with SARS, was a betacoronavirus. It was transmitted from dromedary camels to humans, although like SARS-CoV-1, MERS-CoV was thought to have originated in a bat (perhaps *Neoromicia* sp.) before being transmitted to camels. Although MERS-COV can be transmitted directly from human to human, it was thought that most people who became infected were given the virus from close contact with an infected animal, and it was referred to as a "large zoonotic reservoir."[60] This really highlights why considering animal populations in such epidemics is so important.

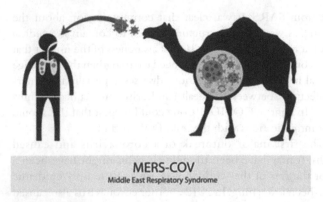

MERS-COV
Middle East Respiratory Syndrome

Figure 1.5 MERS was transmitted to humans from interaction with camels. This poster was made available on Shutterstock by Crystal Eye Studio, but has been made black and white.

Interestingly it appears that the MERS virus used a different entry route into human cells. Instead of using the angiotensin-converting enzyme 2 (ACE2) protein it was thought it used a protein termed DPP4 (CD26). Like SARS-CoV-2 it also seems to involve Transmembrane Protease Serine Type2 (TMPRSS2) and furin,[61] as discussed later.

There were attempts to develop a vaccine against the MERS-CoV virus, and this no doubt paved the way for further development needed when we were hit by SARS-CoV-2.

For further information about MERS, the CRC produced a handy factsheet,[62] whilst others have written reviews in the scientific literature.[63]

Although SARS or MERS were epidemics,[64] neither had the impact of SARS-CoV-2, and the question as to why not, at least for MERS, was addressed by Bauch and Oraby.[65] They discuss an epidemiologist mathematical model, called the basic reproduction number (R_0), which they define as "the average number of infections caused by one infected individual in a fully susceptible population." If R_0 is less than one, the virus will not persist and eventually will die out. However, if R_0 is greater than 1, the spread will exacerbate, and the epidemic will increase. For MERS, the R_0 was estimated to be only 0.69, meaning that the disease would not persist in the human population and eventually the epidemic would fade away. This is what happened in reality.

The R_0 number, called simply R by the media, became something of an obsession during the COVID-19 pandemic on the national news, at least in the United Kingdom (UK). The daily R number was given, as it is affected by human/human contact, which was reduced during lockdowns, and hence the R number should have reduced too. The R was a driver for how long lockdowns would last, and whether limited freedoms would be given, and it also gave an indication of how well the health service may cope with the ongoing pandemic. The R number for COVID-19 was regularly over 1. At the start of the pandemic WHO estimated that the R number was between 1.4 and 2.4,[66] meaning that spread of the virus through the population was likely, and that the pandemic would get worse before it got better. But it fluctuated, at least in the UK, and eventually dropped. A large factor in driving the R number down for SARS-CoV-2 was the availability of the vaccines, as well as an element of herd immunity, i.e., when so many people have been infected there is less likelihood of meeting someone who does not have some immunity. As a comparison, measles, mumps, and chicken pox have R numbers over 10.

Despite the MERS outbreak not being classed as a pandemic, it did spread very widely. One of the most significant outbreaks was in South Korea,[67] a long way from the Middle East where it was most noted. In South Korea at least 24 people died, and there were 166 confirmed cases. The first patient had been to Saudi Arabia, and this is something we need to be cognisant of in the future. Travel around the world is very quick and easy. An airplane can take anyone across to the other side of the globe easily within 24 hours. Nowhere is safe, and if a person is infected then that infection has an ideal way to spread.

Despite SARS and MERS, was the world prepared for COVID-19? It seems not. Saad Ahmed Ali Jadoo certainly did not think so, ending the abstract of their paper by saying:

> It seems that the world was busy preparing for wars and forgot to promote the individual, community, and global health.[68]

1.3.3 Other SARS-like outbreaks

SARS and MERS were the most prominent coronavirus outbreaks, but they were not the only ones. SARS-COV-2 was, in fact, the seventh SARS-like coronavirus[69] to cause disease outbreaks in humans, the others being HcoV-229E, HcoV-OC43, HcoV-NL63, and HcoV-HKU1, as summarised in Table 1.3.

Based on the 10th International Committee on Taxonomy of Viruses (ICTV) report, all human coronaviruses can be classified as:

- Order: *Nidovirales*
- Suborder: *Cornidovirineae*
- Family: *Coronaviridae*
- Subfamily: *Orthocoronavirinae*
- Genera: *Alphacoronavirus*; *Betacoronavirus*; *Gammacoronavirus*; *Deltacoronavirus*

Table 1.3 Known SARS-like outbreaks in the human population

Short name of disease	Type of coronavirus involved	Virus causing disease	Country of origin	Animal of origin/first isolated	Effects seen
COVID-19	*Betacoronavirus*	SARS-CoV-2	China	Bat, through intermediate	An ongoing global pandemic with millions of human deaths
SARS	*Betacoronavirus*	SARS-CoV	China	Bat, through intermediate	Over 8,000 cases, about 700 deaths
MERS	*Betacoronavirus*	MERS-CoV	Saudi Arabia	Bat, through camel	Epidemic with some deaths
	Betacoronavirus	HKU1		Adult with pulmonary disease	Cold-like symptoms
	Alphacoronavirus	NL63		Seven-month-old girl	Cold-like symptoms
	Betacoronavirus	OC43			Cold-like symptoms
	Alphacoronavirus	229E		Probably bat	Cold-like symptoms, but can be more serious
Potential diseases		SARS-like viruses			

However, the COVID-19 pandemic impacts were not restricted to humans. Animals were affected in a range of ways, and the following chapters will discuss some of these issues. In many cases the welfare of animals was significantly reduced, although in some cases animals appeared to fare better.

What is clear is that the COVID-19 pandemic is not over, and the virus may well now circulate in the human population for the foreseeable future, if not forever, in a similar manner to influenzas. Therefore, how such human diseases impact on the welfare of animals, either in captivity or in the wild, needs to be more fully understood and appreciated.

With the prospect of another pandemic there needs to be a plan to prepare for that eventuality. WHO have announced a Pandemic Influenza Preparedness (PIP) Framework,[70] but this has little consideration of animals. There is mention of human–animal interface risk assessments, but as would be expected it is a very anthropocentric view of futureproofing for the emergence of the next virus. What is needed is a more holistic view, embracing our environment and the ecosystems in which we live, and for this we need to consider what happens to the animals. In the following chapters, some of the issues and concerns of animal health, behaviour, and welfare are discussed in light of the COVID-19 pandemic.

Both SARS-CoV and SARS-CoV-2 are *Betacoronaviruses* under the subgenus *Sarbecovirus*. HcoV-OC43 and HcoV-HKU1 are also both *Betacoronaviruses* but in the subgenus *Embecovirus*. HcoV-229E belongs to the genus *Alphacoronavirus*, and it sits in the subgenus *Duvinacovirus*. HcoV-NL63 is also an *Alphacoronavirus*, but it is in the subgenus *Setracovirus*.

HcoV-229E was first isolated in 1966. Like other SARS-like viruses it was thought to originate in a bat, in particular, African hipposiderid bats. The intermediate host was a camelid, so in many respects it had similarities to MERS. The incubation period is relatively short, being only two to five days, and the disease is probably only going to last about 18 days. Symptoms are similar to a common cold, i.e., headache, runny nose, sneezing, and sore throat, but they can be more severe, especially if the patient is immunocompromised. It has also been suggested that this virus may be involved in Kawasaki syndrome. Interestingly it has been suggested that this virus is relatively stable in the environment, and that it can cause outbreaks during the winter.

In 1967 HcoV-OC43 first appeared, or at least was first isolated. Again, it is most prevalent during winter. It also causes cold-like symptoms, including a sore throat. However, it has also been suggested that it can infect neurones, and therefore has the potential to cause encephalitis. Since it first immerged there have been seven genotypes of HcoV-OC43 (termed A-G).

It was not until 2004 that HcoV-NL63 was recognised, in a seven-month-old girl in the Netherlands. Often HcoV-NL63 is involved in co-infections with other viruses, e.g., enterovirus, parainfluenza, or rhinovirus. Symptoms of HcoV-NL63 again are cold-like, including sore throat, cough, and fever. The very young seem to be most susceptible, but all ages may become infected, and, in fact, it has been suggested that annually up to 10% of the population may be affected by HcoV-NL63.

A patient with pulmonary disease in Hong Kong was the first person to be diagnosed as being infected by HcoV-HKU1. Like the other HcoV viruses, symptoms are cold-like, and they include a cough, blocked nose, sore throat, and fever.

With seven known HcoV viruses it is extremely likely that even more will be discovered, or emerge, in the human population in the future. It is not only humans that are vulnerable. Severe acute diarrhoea syndrome coronavirus (SADS-CoV) had a significant effect in pigs, for example.

It is most likely that any future coronavirus will originate in a bat. Bats are known to be efficient reservoirs of coronaviruses, and some survey work has been carried out, especially in China, to try to gauge the potential for viral emergence in future.[71] However, despite such work there are some interesting biological questions that remain a conundrum. Why do bats harbour coronaviruses so readily and why do they not seem to be affected? Why don't they seem to suffer any disease symptoms? And to get from a bat to a human there needs to be cross-species transmission, and what is the route? For some coronaviruses, such as MERS, this has been well established, but we still do not know for SARS-CoV-2, and we may never know. But can we predict what a likely route may be? And lastly, what are the viruses that the bats are harbouring? Should we be worried?

In 2016 a paper reported on work using a metagenomic and reverse engineer approach to gauge the potential for WIV1-CoV SARS-like animal-derived viruses to emerge into the human population. At the end of the "Significance" section the authors conclude:

> Together, the study indicates an ongoing threat posed by WIV1-related viruses and the need for continued study and surveillance.[72]

A year before this the same group published a paper in *Nature Medicine* entitled *A SARS-like cluster of circulating bat coronaviruses shows potential for human emergence*[73]– this was in 2015. In this they concluded that there was indeed a potential risk of a reemergence of a SARS-CoV virus from bats. And they were not the only ones to say such things. Concentrating on the Chinese rufous horseshoe bat (*Rhinolophus sinicus*) and the SARS-like viruses that these animals may harbour, Yuan and colleagues suggested that there was a close evolutionary linkage between the SARS coronaviruses from humans and that seen in this bat species,[74] and they conclude that this enables "defining the scope of surveillance to search for the direct ancestor of human ScoVs." It seems that many people were working on SARS-like viruses with the potential to cause an epidemic/pandemic in humans long before 2019, and many of these researchers were focusing their attention on bats.

It appears that despite the research that pointed to the risk of a bat population allowing the emergence of a SARS virus into the human population the world was unprepared for COVID-19, with many being surprised when it happened.[75] It was known that animals, and in particular bats, harbour a wide range of coronaviruses, and it was known that it was likely that many of these had the potential to infect humans, but the warning was not heeded. What was missing from this conversation seemed to be much discussion about how non-human animals may be affected, and if a virus was going to originate in a bat and infect humans an intermediate animal was likely to be involved. After all, MERS was through a camel, and SARS was known to have an intermediate. In a paper published about ten years after SARS (i.e., 2013), looking at possible strategies to control any future virus emergence, Graham et al.[76] said:

The dispersion of bat species over much of the globe probably enhances their potential to act as reservoirs for pathogens, some of which are extremely virulent and potentially lethal to other animals and humans.

Animals therefore did feature in some discussion. In 2012 a paper suggesting, "A framework for the study of zoonotic disease" took a broad approach, with work which "combines cutting-edge perspectives from both natural and social sciences, linked to policy impacts on public health, land use and conservation."[77] With a focus on possible bat-to-human disease transfer in Ghana a similar paper was published in 2017.[78] However, such discussions did not prevent the SARS-CoV-2 from emerging and causing a pandemic.

Alphacoronviruses have been circulating in the human population for at least twenty years. Viruses such as CoV-HuPn-2018 and HuCCoV_Z1 are associated with acute respiratory illness.[79] Such viruses are similar to those found in a range of animals, including cats, dogs, and pigs, and they have been reported from several countries, including the United States, Thailand, Haiti, and Malaysia. Humans clearly tolerate such viruses circulating freely in populations, but this does not distract from the potential for future pandemics that may lead to human deaths. Vigilance and research on such viruses, where they emanate from, what symptoms they can cause, and how they can be controlled, are needed.

1.4 Effects of SARS-CoV-2 on the human population

The current position with respect to infections and deaths in the human population can be obtained at the WHO dashboard (https://covid19 .who.int/). At the time of writing, 21 July 2023, there were 766,895,075 confirmed cases. Sadly, 6,951,677 people are known to have died. On better news, 13,474,265,907 vaccine doses have been administered (up to 11 July 2023).

However, it was not just the infections and illness that had an effect. One of the biggest effects was forced changes to the social structure of communities. In numerous countries people were told to "social distance," meaning that they were supposed to stay two metres apart at all times if outside their households. Large lines were seen outside shops where entry was restricted to only a limited number of customers at any one time. Supermarkets were operating a one-in/one-out policy. Public transport was limited, and much non-essential work was stopped. Many companies, including our own university, operated a work-from-home policy. This also applied to schools, so classes were being given online. Suddenly, the expectation was that everyone was digitally enabled, having good Wi-Fi and digital tools to support this way of working and studying, which highlighted the phenomenon of "digital poverty."[80] People also had to quickly learn to use new software platforms, including Zoom, Skype, and Teams.

And it became worse with lockdowns, where people were restricted to their homes. For several weeks or months at a time people were told that they could not leave, unless for particular reasons. In the United Kingdom a short walk for exercise was allowed, and essential shopping was allowed. In Spain people could only leave for shopping, and then they had to have proof that this was essential in case the police questioned them. Of course, this was far worse for some people compared to others. Living in a large house with a massive garden gave some freedom. Living in a cramped flat in a tower block was not so great. Anyone who had personal mobility issues, or poor mental health, or was subject to domestic abuse, was even worse off.

And for those who tested positive life was even worse. People were told to isolate within their household, i.e., stay in their bedroom, until they were virus negative. Meals were delivered outside their doors (if they were lucky enough to have someone prepared to do this for them). People were effectively 'imprisoned' for having an infection.

Of course, this was not good for general health. Not surprisingly participation in sport and physical activities declined during COVID-19, and it was suggested that this was particularly significant for older people and, in fact, life expectancy.[81] Other researchers focused on mental health, and, again, it was not a great surprise to find that this also declined in children, adolescents,[82] and adults, both in the UK and in other countries. Those with special needs and or any preexisting mental issues were particularly vulnerable.[83] Women were also reported to be particularly affected.[84] In the United States it was stated:

> As a result of the lockdown measures, the existing gender gap in mental health has increased by 61%. The negative effect on women's mental health cannot be explained by an increase in financial worries or caring responsibilities.[85]

Human health was clearly affected by the pandemic, and not in a good way. As we shall discuss, this was not widely replicated when a range of animals were looked at. For some there was a negative impact, but some animals appeared to fare rather better with less human interference.

1.4.1 Kawasaki syndrome

Some children who have been found to be positive for SARS-CoV-2 present with symptoms of a rare condition known as Kawasaki disease (KD; otherwise known as mucocutaneous lymph node syndrome, which could lead to KD shock syndrome (KDSS)).[86] This disease causes acute vasculitis, i.e., inflammation of blood vessels. Symptoms can include reddened eyes, swollen hands and feet, swollen ("strawberry") tongue, dry and cracked lips, swollen glands, and a rash.[87] Although there seems to be a link in some people to COVID-19,[88] there appears no reports of SARS-CoV-2 causing KD in animals, although there are natural animal models, for example in young dogs,[89] that show similarities to the disease.

1.5 Why it is important to incorporate animals into the discussion: One Health/One Welfare/One Medicine

When we are considering animals, it is worth a quick discussion of what we mean by "animal" and which species are considered. It is easy to see changes in the behaviour and health of mammals, e.g., but what about 'lower' organisms? The simplest animal is *Trichoplax adhaerens*,[90] a rather amorphous creature with little in the way of a body plan. However, like all species, it would have evolved to live in a particular niche in the ecosystem, and as such will contribute to that ecosystem. Also, like all species, it will be influenced by its environment, and as we will see, that environment may have been radically impacted by the pandemic. Therefore, any discussion of spread and mitigating activities of the COVID-19 pandemic needs to consider all animals. Having said that, there is scant information on the impact of COVID-19 for the vast majority of animals worldwide, but here we will discuss some of the more significant and documented impacts seen.

Importantly, we must recognise that animals and humans do not live in isolation from each other, that we all contribute to the vast ecosystem of the Earth. Therefore, even from an anthropomorphic standpoint, we must consider the impact of COVID-19 on the animal populations if we are to truly appreciate its effects on the human race. Additionally, we must acknowledge the impact that the many strategies to mitigate the effects of COVID-19 on humans have had on animals and the implications of this for biodiversity, ecosystem health, and animal welfare.

1.6 Conclusions

The COVID-19 pandemic was, and in many ways continues to be, a significant event for the human population and for society. The governments of many countries imposed restrictions on communities, including social distancing, lockdowns, and self-isolation policies. The corollary of this was the closures of school, many people losing their jobs or being furloughed, and many others being forced to work or study at home. Many companies went out of business and never reemerged as the pandemic eased.

The virus that caused the COVID-19 pandemic almost certainly originated in an animal – a bat (perhaps *Rhinolophus affinis*) – but how it was originally transmitted to the human population has yet to be determined – it may never be known for sure. However, even this suspicion has led to some animal species being susceptible to persecution, whilst others are calling for a One Health policy to prevent this in future as well as for greater efforts to understand the complexities of animal/human interactions and the impacts on both human and animal health and welfare. This will be revisited in the final chapter.

Notes

1 Quammen, D. (2012) *Spillover: The Powerful, Prescient Book that Predicted the Covid-19 Coronavirus Pandemic.* Random House. ISBN-13: 978-0099522850.
2 Excler, J.L., Saville, M., Berkley, S. and Kim, J.H. (2021) Vaccine development for emerging infectious diseases. *Nature Medicine, 27,* 591–600.
3 Frith, J. (2012) The history of plague – Part 1. The three great pandemics. *JMVH, 2,* 2.
4 CDC: The deadliest flu: The complete story of the discovery and reconstruction of the 1918 pandemic virus: https://www.cdc.gov/flu/pandemic-resources/reconstruction-1918-virus.html (Accessed 15/08/23).
5 Nelson, M.I. and Worobey, M. (2018) Origins of the 1918 pandemic: Revisiting the swine "mixing vessel" hypothesis. *American Journal of Epidemiology, 187,* 2498–2502.
6 World Health Organisation: HIV and AIDS: https://www.who.int/news-room/fact-sheets/detail/hiv-aids (Accessed 15/08/23).
7 Reuters: Italy sewage study suggests COVID-19 was there in December 2019: https://www.reuters.com/article/health-coronavirus-italy-sewage-idINL1N2DV2XE (Accessed 12/04/23).
8 Fongaro, G., Stoco, P.H., Souza, D.S.M., Grisard, E.C., Magri, M.E., Rogovski, P., Schörner, M.A., Barazzetti, F.H., Christoff, A.P., de Oliveira, L.F.V. and Bazzo, M.L. (2021) The presence of SARS-CoV-2 RNA in human sewage in Santa Catarina, Brazil, November 2019. *Science of the Total Environment, 778,* 146198.

9 Reuters: Coronavirus traces found in March 2019 sewage sample, Spanish study shows: https://www.reuters.com/article/us-health-coronavirus-spain -science-idUSKBN23X2HQ (Accessed 12/04/23).

10 Zakary, O., Bidah, S., Rachik, M. and Ferjouchia, H. (2020) Mathematical model to estimate and predict the Covid-19 infections in Morocco: Optimal control strategy. *Journal of Applied Mathematics, 2020.* Available here: https:// www.hindawi.com/journals/jam/2020/9813926/ (Accessed 25/02/23).

11 Mahase, E. (2020) Covid-19: First coronavirus was described in The BMJ in 1965. *BMJ*, 369, m1547. https://doi.org/10.1136/bmj.m1547.

12 Almeida, J.D., Berry, D.M., Cunningham, C.H., Hamre, D., Hofstad, M.S., Mallucci, L., McIntosh, K. and Tyrrell, D.A.J. (1968) Coronaviruses. *Nature, 220*, 16.

13 Prasad, S., Potdar, V., Cherian, S., Abraham, P., Basu, A. and ICMR Covid Team (2020) Transmission electron microscopy imaging of SARS-CoV-2. *The Indian Journal of Medical Research, 151*, 241.

14 Das, A., Ahmed, R., Akhtar, S., Begum, K. and Banu, S. (2021) An overview of basic molecular biology of SARS-CoV-2 and current COVID-19 prevention strategies. *Gene Reports, 23*, 101122. https://doi.org/10.1016/j.genrep.2021 .101122.

15 Kb is the abbreviation for kilobase, as in 1,000 bases (nucleotides).

16 Kumar, S., Maurya, V.K., Prasad, A.K., Bhatt, M.L.B. and Sexena, S.K. (2020) Structural, glycosylation and antigenic variation between 2019 novel coronavirus (2019-nCoV) and SARS coronavirus (SARS-CoV). *Virus Disease, 31*, 13–21.

17 Shintake, T. (2020) Possibility of disinfection of SARS-CoV-2 (COVID-19) in human respiratory tract by controlled ethanol vapor inhalation. *arXiv preprint arXiv:2003.12444.*

18 It is technically wrong to consider a virus as alive, although it is a phrase often used with viruses. They have no metabolism and no capability to reproduce without using a host.

19 Gao, Z., Xu, Y., Sun, C., Wang, X., Guo, Y., Qiu, S. and Ma, K. (2021) A systematic review of asymptomatic infections with COVID-19. *Journal of Microbiology, Immunology and Infection, 54*, 12–16.

20 Ledford, H. (2023) Why is COVID life-threatening for some people? Genetics study offers clues. *Nature.* https://doi.org/10.1038/d41586-023 -01655-0 (Accessed 25/05/23).

21 Pairo-Castineira, E., Rawlik, K., Bretherick, A.D., Qi, T., Wu, Y., Nassiri, I., McConkey, G.A., Zechner, M., Klaric, L., Griffiths, F. and Oosthuyzen, W. (2023) GWAS and meta-analysis identifies 49 genetic variants underlying critical COVID-19. *Nature*, May 17, 1–15.

22 Chan, A. and Ridley, M. (2021) *Viral: The Search for the Origin of Covid-19.* Fourth Estate, London.

23 Chan and Ridley (2021).

24 Soon, J.M. and Wahab, I.R.A. (2021) On-site hygiene and biosecurity assessment: A new tool to assess live bird stalls in wet markets. *Food Control, 127*, 108108.

25 Lo, M.Y., Ngan, W.Y., Tsun, S.M., Hsing, H.L., Lau, K.T., Hung, H.P., Chan, S.L., Lai, Y.Y., Yao, Y., Pu, Y. and Habimana, O. (2019) A field study into Hong Kong's wet markets: Raised questions into the hygienic maintenance of meat contact surfaces and the dissemination of microorganisms associated with nosocomial infections. *Frontiers in Microbiology, 10*, 2618.

26 Webster, R.G. (2004) Wet markets—a continuing source of severe acute respiratory syndrome and influenza? *The Lancet, 363*, 234–236.

27 Woo, P.C., Lau, S.K. and Yuen, K.Y. (2006) Infectious diseases emerging from Chinese wet-markets: Zoonotic origins of severe respiratory viral infections. *Current Opinion in Infectious Diseases, 19*, 401.

28 Naguib, M.M., Li, R., Ling, J., Grace, D., Nguyen-Viet, H. and Lindahl, J.F. (2021) Live and wet markets: Food access versus the risk of disease emergence. *Trends in Microbiology, 29*, 573–581.

29 Xiao, X., Newman, C., Buesching, C.D., Macdonald, D.W. and Zhou, Z.M. (2021) Animal sales from Wuhan wet markets immediately prior to the COVID-19 pandemic. *Scientific Reports, 11*, 1–7.

30 Metagenomic: Analysis using sequencing technologies of samples from environmental or clinical origin.

31 GISAID: https://gisaid.org/ (Accessed 12/04/23).

32 Crits-Christoph, A., Gangavarapu, K., Pekar, J.E., Moshiri, N., Singh, R., Levy, J.I., Goldstein, S.A., Suchard, M.A., Popescu, S., Robertson, D.L., Lemey, P., Wertheim, J.O., Garry, R.F., Rasmussen, A.L., Andersen, K.G., Holmes, E.C., Rambaut, A., Worobey, M. and Débarre, F. (2023) Genetic evidence of susceptible wildlife in SARS-CoV-2 positive samples at the Huanan Wholesale Seafood Market, Wuhan: Analysis and interpretation of data released by the Chinese Center for Disease Control. *Zenodo*. https://doi.org/10.5281/zenodo.7754299.

33 *The Atlantic*: The strongest evidence yet that an animal started the pandemic: https://www.theatlantic.com/science/archive/2023/03/covid-origins-research-raccoon-dogs-wuhan-market-lab-leak/673390/ (Accessed 16/05/23).

34 Nature: COVID-origins study links raccoon dogs to Wuhan market: What scientists think: https://www.nature.com/articles/d41586-023-00827-2 (Accessed 16/05/23).

35 Freuling, C.M., Breithaupt, A., Müller, T., Sehl, J., Balkema-Buschmann, A., Rissmann, M., Klein, A., Wylezich, C., Höper, D., Wernike, K. and Aebischer, A. (2020) Susceptibility of raccoon dogs for experimental SARS-CoV-2 infection. *Emerging Infectious Diseases, 26*, 2982.

36 Andersen, K.G., Rambaut, A., Lipkin, W.I., Holmes, E.C. and Garry, R.F. (2020) The proximal origin of SARS-CoV-2. *Nature Medicine, 26*, 450–452.

37 CNN: FBI Director Wray acknowledges bureau assessment that Covid-19 likely resulted from lab incident: https://edition.cnn.com/2023/02/28/politics/wray-fbi-covid-origins-lab-china/index.html (Accessed 01/03/23).

38 Times: What really went on inside the Wuhan lab weeks before Covid erupted: https://www.thetimes.co.uk/article/2e0dcdaa-078c-11ee-9e46-1e1d57315b13?shareToken=1d784fce60f338b19da9d767fe92a553 (Accessed 29/06/23).

39 Guardian: WHO calls on US to share information on Covid-19 origins after China lab claims: https://www.theguardian.com/world/2023/mar/04/who-calls-on-us-to-share-information-on-covid-19-origins-after-china-lab-claims (Accessed 04/03/23).

40 Vox: The coronavirus likely came from China's wet markets. They're reopening anyway: https://www.vox.com/future-perfect/2020/4/15/21219222/coronavirus-china-ban-wet-markets-reopening (Accessed 20/02/23).

41 Lin, B., Dietrich, M.L., Senior, R.A. and Wilcove, D.S. (2021) A better classification of wet markets is key to safeguarding human health and biodiversity. *The Lancet Planetary Health, 5*, e386–e394.

42 Nadimpalli, M.L. and Pickering, A.J. (2020) A call for global monitoring of WASH in wet markets. *The Lancet Planetary Health*, *4*, e439–e440.

43 BBC: Coronavirus: WHO developing guidance on wet markets: https://www.bbc.co.uk/news/science-environment-52369878 (Accessed 20/02/23).

44 Aguirre, A.A., Catherina, R., Frye, H. and Shelley, L. (2020) Illicit wildlife trade, wet markets, and COVID-19: Preventing future pandemics. *World Medical & Health Policy*, *12*, 256–265.

45 Open Access Government: COVID-19 originated at Wuhan wet market via raccoon dogs and foxes: https://www.openaccessgovernment.org/covid-19-originated-at-wuhan-wet-market-via-raccoon-dogs-and-foxes/140578/ (Accessed 20/02/23).

46 Yuan, S., Jiang, S.C. and Li, Z.L. (2020) Analysis of possible intermediate hosts of the new coronavirus SARS-CoV-2. *Frontiers in Veterinary Science*, *7*, 379.

47 Zhao, J., Cui, W. and Tian, B.P. (2020) The potential intermediate hosts for SARS-CoV-2. *Frontiers in Microbiology*, *11*, 580137.

48 CDC: SARS-CoV-2 variant classifications and definitions: https://www.cdc.gov/coronavirus/2019-ncov/variants/variant-classifications.html#anchor_1632158924994 (Accessed 28/02/23).

49 Gong, S.R. and Bao, L.L. (2018) The battle against SARS and MERS coronaviruses: Reservoirs and animal models. *Animal Models and Experimental Medicine*, *1*, 125–133.

50 Lee, S.H. (2003) The SARS epidemic in Hong Kong. *Journal of Epidemiology & Community Health*, *57*, 652–654.

51 WHO: Severe Acute Respiratory Syndrome (SARS): https://www.who.int/health-topics/severe-acute-respiratory-syndrome#tab=tab_1 (Accessed 20/02/23).

52 Feng, D., De Vlas, S.J., Fang, L.Q., Han, X.N., Zhao, W.J., Sheng, S., Yang, H., Jia, Z.W., Richardus, J.H. and Cao, W.C. (2009) The SARS epidemic in mainland China: Bringing together all epidemiological data. *Tropical Medicine & International Health*, *14*, 4–13.

53 Balboni, A., Battilani, M. and Prosperi, S. (2012) The SARS-like coronaviruses: The role of bats and evolutionary relationships with SARS coronavirus. *Microbiologica-Quarterly Journal of Microbiological Sciences*, *35*, 1.

54 Wang, L.F., Shi, Z., Zhang, S., Field, H., Daszak, P. and Eaton, B.T. (2006) Review of bats and SARS. *Emerging Infectious Diseases*, *12*, 1834.

55 Balboni et al. (2012).

56 Feng et al. (2009).

57 Hung, L.S. (2003) The SARS epidemic in Hong Kong: What lessons have we learned? *Journal of the Royal Society of Medicine*, *96*, 374–378.

58 Centers for Disease Control and Prevention (CDC): Middle East Respiratory Syndrome Coronavirus (MERS-CoV): https://www.cdc.gov/coronavirus/mers/index.html (Accessed 22/12/22).

59 WHO: Middle East Respiratory Syndrome Coronavirus (MERS-CoV): https://www.who.int/health-topics/middle-east-respiratory-syndrome-coronavirus-mers#tab=tab_1 (Accessed 22/12/22).

60 Wang, L., Shi, W., Joyce, M. et al. (2015) Evaluation of candidate vaccine approaches for MERS-CoV. *Nat Commun*, *6*, 7712. https://doi.org/10.1038/ncomms8712.

61 Chan, J.F., Lau, S.K., To, K.K., Cheng, V.C., Woo, P.C. and Yuen, K.Y. (2015) Middle East respiratory syndrome coronavirus: Another zoonotic betacoronavirus causing SARS-like disease. *Clinical Microbiology Reviews*, *28*, 465–522.

62 Centers for Disease Control and Prevention (CDC): Middle East Respiratory Syndrome Coronavirus (MERS-CoV) Factsheet: https://www.cdc.gov/coronavirus/mers/downloads/factsheet-mers_en.pdf (Accessed 22/12/22).

63 Chafekar, A. and Fielding, B.C. (2018) MERS-CoV: Understanding the latest human coronavirus threat. *Viruses, 10*, 93.

64 Some referred to SARS as a pandemic: History: SARS pandemic: How the virus spread around the world in 2003: https://www.history.com/news/sars-outbreak-china-lessons (Accessed 25/05/23).

65 Bauch, C.T. and Oraby, T. (2013) Assessing the pandemic potential of MERS-CoV. *The Lancet, 382*, 662–664.

66 Achaiah, N.C., Subbarajasetty, S.B. and Shetty, R.M. (2020) R_0 and R_e of COVID-19: Can we predict when the pandemic outbreak will be contained? *Indian Journal of Critical Care Medicine: Peer-Reviewed, Official Publication of Indian Society of Critical Care Medicine, 24*, 1125.

67 Cowling, B.J., Park, M., Fang, V.J., Wu, P., Leung, G.M. and Wu, J.T. (2015) Preliminary epidemiological assessment of MERS-CoV outbreak in South Korea, May to June 2015. *Eurosurveillance, 20*, 21163.

68 Jadoo, S.A.A. (2020) Was the world ready to face a crisis like COVID-19? *Journal of Ideas in Health, 3*, 123–124.

69 Liu, D.X., Liang, J.Q. and Fung, T.S. (2021) Human coronavirus-229E, -OC43, -NL63, and -HKU1 (*Coronaviridae*). *Encyclopedia of Virology*, 428.

70 WHO: Pandemic Influenza Preparedness (PIP) Framework: https://www.who.int/initiatives/pandemic-influenza-preparedness-framework (Accessed 14/03/23).

71 Fan, Y., Zhao, K., Shi, Z.L. and Zhou, P. (2019) Bat coronaviruses in China. *Viruses, 11*, 210.

72 Menachery, V.D., Yount Jr, B.L., Sims, A.C., Debbink, K., Agnihothram, S.S., Gralinski, L.E., Graham, R.L., Scobey, T., Plante, J.A., Royal, S.R. and Swanstrom, J. (2016) SARS-like WIV1-CoV poised for human emergence. *Proceedings of the National Academy of Sciences, 113*, 3048–3053.

73 Menachery, V.D., Yount, B.L., Debbink, K., Agnihothram, S., Gralinski, L.E., Plante, J.A., Graham, R.L., Scobey, T., Ge, X.Y., Donaldson, E.F. and Randell, S.H. (2015) A SARS-like cluster of circulating bat coronaviruses shows potential for human emergence. *Nature Medicine, 21*, 1508–1513.

74 Yuan, J., Hon, C.C., Li, Y., Wang, D., Xu, G., Zhang, H., Zhou, P., Poon, L.L., Lam, T.T.Y., Leung, F.C.C. and Shi, Z. (2010) Intraspecies diversity of SARS-like coronaviruses in *Rhinolophus sinicus* and its implications for the origin of SARS coronaviruses in humans. *Journal of General Virology, 91*, 1058–1062.

75 *The Globe and Mail*: Making history: How a pandemic took the world by surprise: https://www.theglobeandmail.com/opinion/article-making-his-tory-how-a-pandemic-took-the-world-by-surprise/ (Accessed 24/02/23).

76 Graham, R.L., Donaldson, E.F. and Baric, R.S. (2013) A decade after SARS: Strategies for controlling emerging coronaviruses. *Nature Reviews Microbiology, 11*, 836–848.

77 Wood, J.L., Leach, M., Waldman, L., MacGregor, H., Fooks, A.R., Jones, K.E., Restif, O., Dechmann, D., Hayman, D.T., Baker, K.S. and Peel, A.J. (2012) A framework for the study of zoonotic disease emergence and its drivers: Spillover of bat pathogens as a case study. *Philosophical Transactions of the Royal Society B: Biological Sciences, 367*, 2881–2892.

78 Lawson, E.T., Ohemeng, F., Ayivor, J., Leach, M., Waldman, L. and Ntiamoa-Baidu, Y. (2017) Understanding framings and perceptions of spillover:

Preventing future outbreaks of bat-borne zoonoses. *Disaster Prevention and Management*, *26*, 396–411. https://doi.org/10.1108/DPM-04-2016-0082.

79 Vlasova, A.N., Toh, T.H., Lee, J.S.Y., Poovorawan, Y., Davis, P., Azevedo, M.S., Lednicky, J.A., Saif, L.J. and Gray, G.C. (2022) Animal alphacoronaviruses found in human patients with acute respiratory illness in different countries. *Emerging Microbes & Infections*, *11*, 699–702.

80 Seah, K.M. (2020) COVID-19: Exposing digital poverty in a pandemic. *International Journal of Surgery (London, England)*, *79*, 127.

81 Harangi-Rákos, M., Pfau, C., Bácsné Bába, É., Bács, B.A. and Kőmíves, P.M. (2022) Lockdowns and physical activities: Sports in the time of COVID. *International Journal of Environmental Research and Public Health*, *19*, 2175.

82 Singh, S., Roy, D., Sinha, K., Parveen, S., Sharma, G. and Joshi, G. (2020) Impact of COVID-19 and lockdown on mental health of children and adolescents: A narrative review with recommendations. *Psychiatry Research*, *293*, 113429.

83 Panchal, U., Salazar de Pablo, G., Franco, M., Moreno, C., Parellada, M., Arango, C. and Fusar-Poli, P. (2021) The impact of COVID-19 lockdown on child and adolescent mental health: Systematic review. *European Child & Adolescent Psychiatry*, 1–27.

84 Banks, J. and Xu, X. (2020) The mental health effects of the first two months of lockdown during the COVID-19 pandemic in the UK. *Fiscal Studies*, *41*, 685–708.

85 Adams-Prassl, A., Boneva, T., Golin, M. and Rauh, C. (2020) The impact of the coronavirus lockdown on mental health: Evidence from the US. https://doi.org/10.17863/CAM.81910.

86 Whittaker, E., Bamford, A., Kenny, J., Kaforou, M., Jones, C.E., Shah, P., Ramnarayan, P., Fraisse, A., Miller, O., Davies, P. and Kucera, F. (2020) Clinical characteristics of 58 children with a pediatric inflammatory multisystem syndrome temporally associated with SARS-CoV-2. *JAMA*, *324*, 259–269.

87 NHS: Kawasaki disease: https://www.nhs.uk/conditions/kawasaki-disease/ (Accessed 22/02/23).

88 Viner, R.M. and Whittaker, E. (2020) Kawasaki-like disease: Emerging complication during the COVID-19 pandemic. *The Lancet*, *395*, 1741–1743.

89 Felsburg, P.J., HogenEsch, H., Somberg, R.L., Snyder, P.W. and Glickman, L.T. (1992) Immunologic abnormalities in canine juvenile polyarteritis syndrome: A naturally occurring animal model of Kawasaki disease. *Clinical Immunology and Immunopathology*, *65*, 110–118.

90 Scientific American: https://www.scientificamerican.com/article/worlds-simplest-animal-reveals-hidden-diversity/ (Accessed 12/10/22).

2

Can the susceptibility of animals to SARS-CoV-2 be predicted?

2.1 Introduction

The tenet of this book is that the COVID-19 pandemic has had effects on animals. Such effects may be indirect, where human action has an impact on animal welfare, and these will be discussed in subsequent chapters. However, sometimes we forget that humans are animals, and as such we are not much different from many other animal species. Biochemically, different animal species can be very similar, even if they do not look the same. With this in mind, it needs to be remembered that the infection of cells by a virus such as SARS-CoV-2 is dictated by the interaction of the virus with the target cell, with the subsequent injection of the viral genetic material (nucleic acids: RNA or DNA) into the cell. Once inside, the genetic material can be used to force the cell to make new viral particles, which can be released to sustain the viral "life cycle,"[1] and used to infect the next host. Hence, diseases such as COVID-19 can be propagated in an individual and across populations.

All the processes of host recognition, interaction, and subterfuge are biochemical events. The recognition of the cell by the virus relies on the virus being able to interact with the surface properties of the cells, and this is dictated by the proteins' presence (which may be glycosylated with sugars) that are embedded in a lipid membrane. This is a complex and convoluted surface, acting as a barrier to entry to anything that the cells do not desire, such as a harmful virus. Therefore, the virus has to be able to recognise that it is the right cell to attack, and then have a strategy for getting in. A good analogy would be a warring army trying to invade a medieval castle.[2] The commander needs to know which castle to attack,

DOI: 10.1201/9781003427254-2

and then how to gain entry, so they can hijack the resources there for their own use, making the army sustainable to invade the next castle, and the next, and so on.

Just like the castle analogy, the cell does not simply allow this invasion to succeed, but the cells, with the aid of other cells in the body, will wage a defence. Hence, in humans not only does the cell have a physical barrier, but also its compatriots will mount an immune response. This is what goes awry as the cytokine storm, which can end in death. Inside the cell, things are not quiescent either, as the cell will mount a personal stress response, which may involve an increase in the accumulation of reactive oxygen species, i.e., small reactive molecules containing oxygen (a process often dubbed as oxidative stress), for example. Therefore, the virus does not have an easy ride.

In humans, at least for some, the SARS-CoV-2 virus has evolved the correct strategy and cellular invasion is successful. The level of disease does vary from individuals being asymptomatic, or having a nasty few days of disease, or death. It is hard to predict for an individual human what that outcome may be. Gallo Marin et al.[3] state that indicative factors include "demographic, clinical, immunologic, hematological, biochemical, and radiographic findings." Therefore, it is a complex landscape.

If it is difficult for a species where we know the disease takes place, i.e., humans, what is the chance of predicting this for animal species? Here we explore some of the evidence and ideas.

2.2 A quick crash course in molecular biology

Molecular biology can be a little daunting if you are not familiar with the terminology, so here is a quick overview. Of course, such materials can also be found readily on the Internet.[4]

The molecular components that carry out the actions inside cells are the proteins. A protein may carry out a chemical reaction, and, if it does, it is referred to as an enzyme. Alternatively, proteins may be structural, holding the other cellular components in the right manner. Either way, proteins are instrumental to how a cell functions and survives.

A protein is composed of a string of amino acids. Amino acids are chemical units that can be combined together to make a larger structure, a little like Lego™ bricks. There are twenty naturally occurring amino acids. They all have a common core structure. They have a "nitrogen" end, which is a chemical group called an amino group ($-NH_2$), while the other end is the carbon end, actually a carboxyl ($-COOH$) group. The amino group of one amino acid can chemically bind with the carboxyl group of another amino acid, creating a dipeptide (i.e., two amino acids joined together). However, the process can be repeated, with the "free" end of the dipeptide binding to another amino acid, and then another, and so a string of amino acids can be added in a long line. This is a polypeptide, or protein. It is a little like a long line of beads on a string. However, each amino acid has a central carbon atom. Carbon can bond to four things at the same time, so it has an amino group and carboxyl group taking two of these opportunities. The third binding is always to a proton (H). That leaves the fourth bond, and this is to variable chemical groups, the so-called side group, or R. Each of the twenty natural amino acids has a different side group, and this gives the amino acid its unique characteristic. What is important here is that once the amino acids are joined as a string, the side groups are then in an order too, depending on which amino acid has been used at the point in the string. Therefore, proteins can be considered as a long line of side groups, each with its own chemical characteristic (Figure 2.1). Some side groups are inert (e.g., glycine only has a proton: H), others may be acid, or sulphur containing[5] (e.g., cysteine has a -SH group) etc. Where such chemical groups are in a protein will bestow on that protein its function and activity.

However, proteins are not a simple line of amino acids like an unravelled piece of string. Proteins have a three-dimensional structure. Some amino acids are water-hating (hydrophobic), while others are water-loving (hydrophilic). Imagine a polypeptide being dropped into water. All the water-hating side groups would wish to hide, while the water-loving ones would try to swim. This is what generally happens. The protein folds so that the hydrophobic side groups are sequestered inside a ball-like structure, while the water-loving ones will line the surface of that ball. Therefore, proteins will have a defined structure that is dependent on the amino acids from which it is composed. The proteins of the virus are no different, and they will therefore have a defined structure (shape). This is seen, e.g., with the spike proteins.

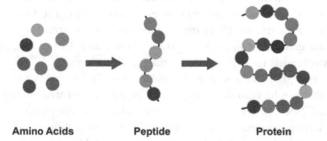

Amino Acids   **Peptide**   **Protein**

Figure 2.1 The basic units of proteins are amino acids, which can be joined to make a peptide, and this process can be continued to create larger structures, the proteins. Image from Shutterstock courtesy of Ali DM (made black and white).

The vast majority of a protein has little to do but to maintain the three-dimensional structure of that protein, and often to let it move and flex in defined ways. If the activity of a protein is to bond two other chemicals together, perhaps in a metabolic pathway, amino acids must be on the surface to interact with those chemicals, and this region of the protein is referred to as the active site. However, an active site may only comprise a handful of amino acids, while the rest of the protein is merely there to make sure that those few amino acids are in the correct three-dimensional orientation, and proteins can be hundreds of amino acids. Therefore, when studying proteins, it is often worth focussing on just a few amino acids, as will be seen here later.

So, what determines the order of the amino acids in a protein?

Most cells contain deoxyribonucleic acid (DNA). This, too, is a string of Lego™-like components, but here the units are nucleotides,[6] not amino acids. But in a similar manner, the nucleotides have side groups, which are called bases. Unlike amino acids, there are only four types of base: adenine (A), thymine (T), guanine (G), and cytosine (C). It is the order of these bases along the piece of string of DNA that makes up the code embedded in the DNA, as illustrated in Figure 2.2.

It was Watson and Crick,[7] using x-ray data from Rosalind Elsie Franklin (1920–1958),[8] that determined that DNA was, in fact, a double helix, with the bases "holding hands" across the two strands of the helix.

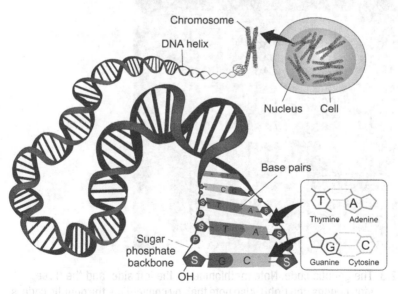

Figure 2.2 DNA is composed of four nucleotides, each with a different base, and it is the order of these that comprises the genetic code. Image from Shutterstock courtesy of Soleil Nordic, but made black and white.

What concerns us here is that it is the order of bases that is important. However, there are only four and there are twenty amino acids, so how can only four bases encode for so many amino acids? The answer to this conundrum comes from the fact that the bases are "read off" in groups of threes, the so-called codons. However, there are 64 ways of grouping four different things as threes ($4^3 = 64$). Within these 64 combinations of codons, there is one that encodes for start (ATG), and this also encodes for (tells a cell how to make a specific protein) the amino acid methionine, so, by default, all proteins start with this amino acid (although it can be removed). There are three codons that indicate for the process to stop. That leaves 60 codons. It would be logical that there are therefore three codons for each amino acid, but this is not the case. As can be seen in Figure 2.3, some amino acids are encoded for by more, such as serine, which has six codons.

The sequence of codons, from the start to the stop, constitutes what is referred to as an "open reading frame" (ORF) and this is the part of a gene that encodes for the protein sequence. Of course, the gene for the proteins needs to have its expression controlled, but such mechanisms are beyond the scope of this brief discussion.

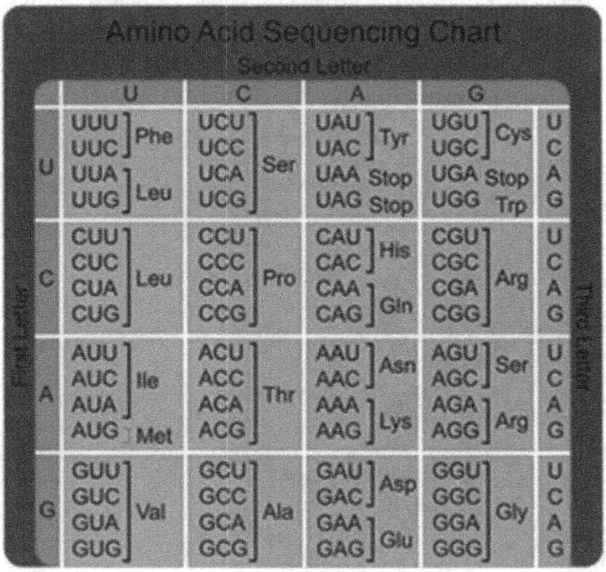

Figure 2.3 The genetic code. Note methionine on the left side, and the three stop codons (top right). Also note that, by convention, the genetic code is shown as RNA, in which T (thymine) is substituted by U (uracil). Image from Shutterstock, courtesy of gstraub, but made black and white.

The important points here are that the order of the amino acids in a protein is important for its structure and therefore function, and these can be listed as a string of code. All amino acids have a one-letter code, so we can represent them as a line of letters, such as AGYM, etc., which translates to alanine, followed by glycine, then tyrosine, and then methionine. All proteins can be represented in this way, in what is referred to as a primary sequence. This is what is used in the discussion in the sections below. However, the order of the amino acids can be determined easily from the DNA sequence, and this can be used for analysis too, although it is not used here.

Second, protein structure is vital. In the discussion below, proteins on the virus need to be "recognised" by the proteins in the host cell, or else the virus has no way to invade the cells. Therefore, each of the proteins involved has a unique structure, and they are able to fit together, a little like two people shaking hands. These structures are determined by the amino acids, and so by looking at the sequences we can get, perhaps a little naïvely, an indication of whether the structures are likely to be correct and therefore work.

Proteins are not necessarily found as they are originally made. Some are cleaved before they become active. This may allow a new structure to be formed, or simply reveal a structure previously hidden. The cleavage events are carried out by other proteins, so-called proteases (or peptidases). A classic example of this is the hormone insulin. This was one of the proteins worked on by the brilliant Dorothy Crowfoot Hodgkin (1910–1994)[9] at the University of Oxford. Insulin is two different polypeptides (proteins) held together by sulphur–sulphur binding, but despite the fact that it is two proteins, there is only one gene that encodes it. It is made as one protein, which is then cleaved at three places and creates the final structure. Without these cleavage events, insulin would not have the correct structure and hence would not work. The process of SARS-CoV-2 invasion of cells involves proteins that have such activity of cleaving others, i.e., peptidases.

Having had a whistle-stop tour through molecular biology, can any of this help us determine if an animal is susceptible to the SARS-CoV-2 virus?

The full complement of genes in an organism is referred to as the genome. In humans, in the genome there are a little over 20,000 protein-coding genes, although the exact number has been a little debatable.[10] However, humans probably have over 100,000 different proteins, although again the exact number is not yet known.[11] Therefore, the first question to be determined when considering how SARS-CoV-2 interacts with a host is which of the host proteins are involved? In the first instance, most work has concentrated on just one, referred to as ACE2

(Angiotensin–converting enzyme 2), and it is here that we will start our discussion below.

The genome of the SARS-CoV-2 is a little different from that of humans, and it is, in fact, very complex.[12] The genome is comprised of ribonucleic acid (RNA) rather than DNA. The main difference is that the backbone of the structure contains ribose rather than deoxyribose. The general principles of action remain the same. The viral genome is 29,903 nucleotides in length and has two major open reading frames, ORF1a and ORF1b, which are used to produce numerous proteins, partly through the action of polypeptide cleavage. The genome also has a series of what is referred to as subgenomic mRNAs, and it is one of these, S, that encodes for the spike protein. It is this spike protein that allows the virus to interact with the host cells. In fact, it is only a small section of the spike protein that is particularly needed for this action, a region known as the receptor-binding domain (RBD), which is part of what is known as subunit S1, and comprises amino acids 319 to 541.[13] Therefore, something on the host cells needs to recognise this RBD and allow the virus to "stick," and it has been an interaction between the viral RBD and human ACE2 that has been a particular focus of research.

2.3 Can molecular biology help?

The first indication of the interaction of the virus with host cells suggested that the most important host protein was Angiotensin-converting enzyme 2 (ACE2). ACE2 is an enzyme that has peptidase activity, i.e., its activity is to break down other proteins. It is this protein that is thought to be recognised by the spike proteins, which are protruding from the virus. This allows the virus to effectively stick to the cell surface, which would be the first prerequisite for viral entry. Therefore, if the interaction between ACE2 and spike protein can be characterised, there would be a better understanding of how the virus works, with the corollary that we could then do something about it.

ACE2 itself is made up of 805 amino acids in humans. Of course, it will be a small subset of these that are involved in the protein's interaction with the spike proteins. The challenge is to understand which of the amino acids are used. The protein will adopt a complex three-dimensional structure, so it is not even necessarily contiguous amino acids that are involved.

As can be seen in Figure 2.4,[14] the proteins are large and complex, and only small patches of each are involved in the interactions. Several groups have worked on obtaining a better understanding of this molecular event.[15] It is therefore theoretically possible to pin down the amino acids of importance and to disregard the rest.

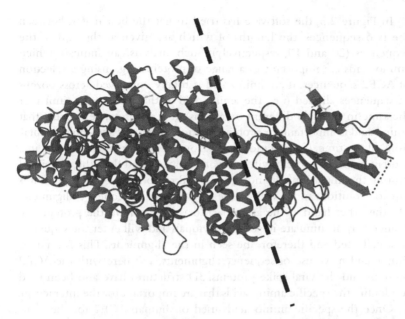

Figure 2.4 The crystal structure of chimeric omicron RBD (strain BA.1) complexed with human ACE2. The ACE2 protein is the structure on the left (original in green), and the spike protein is on the right (original in purple), with a black dotted line indicating the divide. Database entry by Geng, Q., Shi, K., Ye, G., Zhang, W., Aihara, H. & Li, F. in March 2022.

The question is: does the same happen in any animals? First, this would rely on the presence of the ACE2 protein in that animal. Therefore, the DNA encoding that protein must exist in the animal's genome.[16]

It is relatively easy to align both DNA and amino acid sequences using bioinformatic tools, such as *Clustal Omega*.[17] Creating two random amino acid sequences and running it through the algorithm gives the output seen in Figure 2.5.

```
Test1      AGIVVTYGSCT--YKIPVAGTQ  20

Test2      -AIIVGTYSCMPTYGAPMPV--  19

           .*:*    **   *  *:
```

Figure 2.5 A typical output from the alignment of two random sequences using *Clustal Omega* (run in November 2022). Hyphens inserted into the sequence are gaps added automatically, although this can be turned off. Under the sequences are similarity scores, based on the chemical composition of the amino acid side group: * indicates a perfect match, : indicates a conserved change, . is a less well-conserved change, whilst no indication means there is no similarity.

In Figure 2.5, the software has tried to get the best match between the two sequences, the lengths of which are given at the end of the sequences (20 and 19, respectively). Such analysis can indicate which amino acids are important in a range of proteins; e.g., using a selection of ACE2 sequences, if an amino acid is totally conserved across dozens of sequences aligned (i.e., the amino acid at that position is similar or the same in multiple different organisms), it is a reasonable bet that that amino acid is important for protein function. Genomes accumulate mutations all the time. If a mutation knocks out an important amino acid in a protein, then the protein will not work properly and then the individual hosting that mutation may not survive and reproduce, so the mutation is lost to evolution, and it will not be revealed in any sequence alignment. On the other hand, if the amino acid has little role in the protein, the change may accumulate in some individuals and will enter the sequence data collected and therefore be seen in the alignments. This is a rather simple and naïve use of sequence alignments, and here with the ACE2 proteins and the viral spike proteins, 3D structures have also been used to identify the specific amino acids that are important for the interaction.

Once the specific amino acids used on human ACE2 for the virus activity have been identified, one can look in other animal species to see if they are present. If they are, it is a good indicator that, at least for that phase of the viral life cycle, the virus can interact with the host cells and start the entry process. This is no guarantee that disease will ensue, but if the amino acids are absent then it is a good indicator that viral entry would be blocked, or at least it would need a different mechanism.

A literature search can uncover the amino acids in ACE2 researchers think are important in humans, and therefore these can be searched for in a variety of animal species. The amino acids can be identified using papers by Sun et al.,[18] Damas et al.,[19] Wan et al.,[20] etc. An example is given in Figure 2.6, where part of the human ACE2 sequence is aligned with the ACE2 proteins from eleven vertebrate species. These include a bat species, particularly pertinent as the SARS-CoV-2 virus probably has a bat origin. Non-human primates are included, as are rodents, companion animals (dogs and cats), camels (source of MERS), and a representative marine mammal.

Ignoring the non-consequential amino acids, it can be seen that non-human primates are very likely to be susceptible to the SARS-CoV-2 virus. As we will discuss in Chapter 3, this has been borne out in reality. On the other side of the coin, the amino acids in the significant positions of the ACE2 proteins in fish and birds are more variable, and it appears, from this naïve analysis, that these animals may be safe from the virus.

It can be concluded, from this simple discussion, that molecular biology does have the potential to help in identifying animals that may be vulnerable, and therefore their welfare during the pandemic has to have special measures, perhaps. Obviously, real-life reports of infected animals will need to be taken into account too.

Output from *Clustal Omega*, including consensus analysis as the last row:[17]

```
Fish     MSTA--GRV-AAGAAAMLLLVVALLTPGLRAQVDTETRARAFLEKFSTEASVKMYDYSLA   57
Bird     --------MDMLVCIWLLCG----LIAVVSPQT-VTQQAQMFLEEFNKRAEDINYESSLA   47
Bat      ----------MSGSSWLFLS----LVAVAAAQSSTEEKAKIFLENFNSKAEDLSHESALA   46
Rodent   MGSCPGARGKMLGSSWLLLS----FVAVTAAQSTIEELAKTFLDKFNQEAEDLDYQRSLA   56
Rabbit   ----------MSGSSWLLLS----LVAVTAAQSTIEELAKTFLEKFNQEAEDLSYQSALA   46
Macaque  ----------MSGSSWLLLS----LVAVTAAQSTIEEQAKTFLDKFNHEAEDLFYQSSLA   46
Human    ----------MSSSSWLLLS----LVAVTAAQSTIEEQAKTFLDKFNHEAEDLFYQSSLA   46
Gorilla  ----------MSGSSWLLLS----LVAVTAAQSTIEEQAKTFLDKFNHEAEDLFYQSSLA   46
Cat      ----------MSGSFWLLLS----FAALTAAQSTTEELAKTFLEKFNHEAEELSYQSSLA   46
Dog      ----------MSGSSWLLPS----LAALTAAQS-TEDLVKTFLEKFNYEAEELSYQSSLA   45
Camel    ----------MSGSSWLLLS----LVAVTAAQSTTEELAKTFLEEFNHEAADLSYQSSLA   46
Dolphin  ----------MSGSFWLLLS----LVAVTAAQSATEERAKTFLQKFDREAEDLSYQSSLA   46
                   .  ::        :      *      .: **::*. .*.   :: :**
```

Figure 2.6 Alignment of sequences of ACE2 receptor proteins for 12 animal species, Highlighted (vertical) the amino acids used for SARS-CoV-2 binding.

Amino acids that have been highlighted vertically are suggested as being important by Shang et al. (2020),[15] Sun et al. (2020),[18] and Damas et al. (2020).[19] Horizontal grey highlighting shows the human sequence for clarity. Animal sequences chosen were representative examples of animal groups. Specifically, they were: Fish: cod (*Gadus morhua*) Ac: XP_030232530.1; Bird, Rock Pigeon (*Columba livia*) Ac: XP_021154486.1; Bat (*Myotis lucifugus*) Ac: XP_023609437.1; Rodent (*Ictidomys tridecemlineatus*) Ac: XP_005316051.3; Rabbit (*Oryctolagus cuniculus*) Ac: QHX39726.1; Macaque (*Macaca nemestrina*) Ac: XP_011733505.1; Humans (*Homo sapiens*) Ac: NP_001358344.1; Gorilla (*Gorilla gorilla gorilla*) Ac: XP_018874749.1; Cat (*Lynx pardinus*) Ac: VFV30336.1; Dog (*Canis lupus familiaris*) Ac: QJS40032.1; Camel (*Camelus ferus*) Ac: XP_006194263.1; Dolphin; Common bottlenose (*Tursiops truncates*) Ac: XP_019781177.2.[21]

2.3.1 The use of bioinformatics to determine animal susceptibility, which proteins should be looked at, and why

From the discussion above, the ACE2 protein is the most obvious target for bioinformatic analysis, and this has been used by others.[19(Damas),22] However, it is not the only protein of interest. There are numerous proteins involved in the viral entry and then the progression of the virus life cycle, as well as the numerous proteins that are used as part of the host defence system.

With this in mind, we took the approach outlined above to look at three other proteins.[23] Transmembrane Protease Serine Type2

(TMPRSS2), like ACE2, is a protease, and it has been found to be used in viral entry.[24] As with ACE2, the protein amino acid sequence for this protein in a range of animals could be sourced from the NCBI database.[25] Our conclusion was that "TMPRSS2 may be helpful in determining susceptibility," but this helpfulness was very limited and could not be used in isolation. Two other proteins studied were furin and neuropilin-1. Furin is also a protease, and it has been suggested as a therapeutic target for COVID-19,[26] and is known to be involved in the entry of viruses into host cells, such as Ebola.[27] Neuropilin-1 is a cell surface receptor that is also thought to be important in the SARS-CoV-2/host interaction.[28] When these two proteins were looked at with the view towards susceptibility of the SARS-CoV-2 virus, we stated, "whereas the other two [furin and neuropilin] would appear to be of limited use." Little consensus was found in the analysis that would give much optimism for pursuing work on furin or neuropilin-1 in the future.

One of the aspects of the genetic information that is of possible use is the fact that individuals have polymorphisms[29] in their genes. These are often single-base changes in the DNA sequences, which are known as single nucleotide polymorphisms (SNPs, pronounced as snips). The presence of SNPs in individuals can be correlated with the severity of disease, and this can be carried out for COVID-19.[30] Therefore, as well as conserved amino acids in the proteins, significant changes to those amino acids can be analysed too to see if there is any indication of increased or decreased disease severity in either individual humans or animal species. As ACE2 appeared to have limited predictive value (see below), polymorphism analysis for ACE2, TMPRSS2, furin, and neuropilin-1 was carried out, but the data obtained were of limited use for identifying animal species at risk. However, as more polymorphisms are identified, it may be a more fruitful approach in the future. After all, there are numerous cell surface proteins and potentially thousands of polymorphisms that may be helpful for this type of analysis in the future.[31]

However, it is not only cell surface proteins that may have an influence on disease progression. One protein that was highlighted in the literature was glutathione S-transferase-omega. Glutathione is an immensely important small peptide molecule found in cells where it plays an instrumental role in controlling the intracellular redox (reduction/oxidation, the measure of the free-radical levels and other oxidants in the cell) state of the cell. Lack of control of cellular oxidation states can lead to a situation called oxidative stress,[32] and ultimately it can lead to cell death. One of the reasons for eating fruit and vegetables is to aid in the control of this mechanism.[33]

Glutathione exists in two forms: the reduced, GSH, and the oxidised, GSSG – two GSH molecules covalently joined across a disulphide bridge.[34] Cells use a lot of energy to maintain a high concentration of GSH and a relatively low concentration of GSSG, and this keeps the redox state very reduced. The reduced form, GSH, is readily oxidised into GSSG by reactive oxygen species and other free radicals; thus, GSH can be thought of as a "free radical mopper," reducing the levels of the toxic chemicals. This is important, as cells rely on reduced compounds such as nicotinamide adenine dinucleotide (NADH) and the phosphorylated form (NADPH) for much of their metabolism, including the formation of adenosine triphosphate (ATP), which is used as an energy source in cells.[35]

In mammals, there are eight classes of glutathione S-transferases, but it is the omega form (GSTO) that is relevant here. These enzymes are involved in maintaining GSH homeostasis, but they have also been found to have a role in chronic obstructive pulmonary disease (COPD), suggesting a relevance to COVID-19. One isoform, GSTO1, activity also impacts the inflammatory response, again relevant to a SARS-Cov-2 infection. GSTO2 is also involved in redox homeostasis and control of vitamin C levels, the latter of which has been shown to be helpful for COVID-19 patients. It has been noted that polymorphisms in the GSTO1 and GSTO2 genes may correlate with COVID-19 severity in humans.[36] The authors state that "individuals carrying variant GSTO1*AA and variant GSTO2*GG genotypes exhibit higher odds of COVID-19 development, contrary to ones carrying referent alleles." Of significance is their following statement that "Carriers of H2 haplotype, comprising GSTO1*A and GSTO2*G variant alleles, were at 2-fold increased risk of COVID-19 development." Therefore, here is an enzyme downstream[37] of the initial virus attack on the surface of the host's cells, and variation of the amino acid composition of those enzymes, and therefore one assumes their activity, has a profound effect on the development and severity of disease. The question we asked was: would such polymorphisms exist in animals and does that correlate to the disease in animals that has been reported? So we had a look.[38]

GSTO1 SNP (rs4925[39]) leads to a nucleotide change of C to A, and this translates to an amino change of A140D (Ala to Asp). Therefore, GSTO1 was aligned with homologous sequences from 25 different vertebrate species, from fish to gorillas, that covered this area of the gene. GSTO2 SNP (rs156697) is at amino acid 142 and this leads to changes of N142D or N142Y. Here this was aligned with fifteen animal sequences. The outcome of this rather limited analysis was that GSTO1 SNPs gave no indication of disease susceptibility, but the presence of the GSTO2 SNP did seem to correlate with a species' susceptibility for SARS-CoV-2 in some cases, and we suggested that this ought to be investigated further.

Therefore, in theory at least, there are some genes beyond ACE2 that, when analysed in more detail, may give some indication of susceptibility to COVID-19 in animals. So, what did other researchers find?

2.4 Which animals are predicted to be infected with SARS-CoV-2?

Work such as that discussed above needs to build on, and be compared to, the published studies of others. Do we all find the same thing, and were the data useful?

As the pandemic unfolded, there were a flurry of papers where the amino acid sequences of interacting proteins were looked at in a range of animals. Most of these concentrated on the ACE2 protein, as this was very early on found to be important for facilitating human infections. One of the largest studies was carried out by Damas et al. Using the ACE2 protein, some structural information, and noting the amino acids of most relevance in ACE2 function, they looked across the amino acid sequences of over 400 vertebrate species. They then characterised the animals into groups, depending on their predicted susceptibility to SARS-CoV-2: very high, high, medium, low, very low. Some of their predictions, ones we thought of note, are listed in Table 2.1.

A look through Table 2.1 gives some data, which is expected, but also throws up a series of surprises. The non-human primates come out as highly susceptible, but as in some cases the ACE2 protein is exactly the same as humans, it is not unexpected. However, SARS-CoV-2 was thought to have originated in a bat (*Rhinolophus affinis*)[41] and yet there are several bat species in the very low susceptibility bracket. Of course, not all bats are the same; it being a wide range of animals belonging to the order Chiroptera, with an estimated 1,240 species of bat in the world, so it would be naïve to group them together, and, in fact, they are divided into Megachiroptera (megabats) and Microchiroptera (microbats).[42] Even so, having so many found in the very low susceptible group by the analysis of Damas et al. may look odd. Many people suspect that the SARS-CoV-2 virus was transmitted from bats to humans through a pangolin, and yet pangolins are predicted to have very low susceptibility here (Figure 2.7).

MERS was thought to have emanated from a camel, as discussed above, and yet camels are predicted to have low susceptibility to SARS-CoV-2, a virus closely related to MERS. Again, this seems a little odd, perhaps.

Here, some species relevant to our later discussions are also picked out of the data Damas et al. presented. Many feline species are in the medium group, while dogs are in the low susceptibility class. Notable here is that minks and ferrets (*Mustela*) are in the very low group, and yet, as discussed later, infection of minks has been a major problem, leading to the slaughter of millions of animals. Therefore, perhaps the use of ACE2 as a marker to predict an animal's susceptibility to SARS-CoV-2 is not effective, despite the massive effort Damas and her group put into this work. This highlights the need for

Table 2.1 Some notable predictions of animals susceptible to SARS-CoV-2

Animal	Latin name	Prediction of susceptibility	On IUCN red list
Gorilla (Western lowland)	*Gorilla gorilla gorilla*	Very high	Yes
Macaque (Crab-eating)	*Macaca fascicularis*	Very high	No
Macaque (Southern pig-tailed)	*Macaca nemestrina*	Very high	Yes
Chimpanzee	*Pan troglodytes*	Very high	Yes
Bonobo	*Pan paniscus*	Very high	Yes
Baboon (Olive)	*Papio anubis*	Very high	No
Rat (Gambian pouched)	*Cricetomys gambianus*	High	No
Hamster (Chinese)	*Cricetulus griseus*	High	No
Whale (Beluga)	*Delphinapterus leucas*	High	No
Orca	*Orcinus orca*	High	Yes
Whale (Minke)	*Balaenoptera acutorostrata scammony*	High	No
Porpoise (Harbour)	*Phocoena phocoena*	High	No
Dolphin (Common bottlenose)	*Tursiops truncates*	High	Yes
Narwhal	*Monodon monocerus*	High	No
Lemur (Blue-eyed black)	*Eulemur flavifrons*	High	Yes
Deer (White-tailed)	*Odocoileus virginianus texanus*	High	No
Deer (Pere David's)	*Elaphurus davidianus*	High	Yes
Muskrat	*Ondatra zibethicus*	High	Yes
Anteater (Giant)	*Myrmecophaga tridactyla*	High	Yes
Aye-aye	*Daubentonia madagascariensis*	Medium	Yes
Hamster (Golden)	*Mesocricetus auratus*	Medium	Yes
Whale (Sperm)	*Physeter catodon*	Medium	Yes

(*Continued*)

43

Table 2.1 (Continued) Some notable predictions of animals susceptible to SARS-CoV-2

Animal	Latin name	Prediction of susceptibility	On IUCN red list
Squirrel (Daurian ground)	*Spermophilus dauricus*	Medium	No
Bison (American)	*Bison bison bison*	Medium	Yes
Yak (wild)	*Bos mutus*	Medium	Yes
Cattle	*Bos taurus*	Medium	No
Buffalo (Water)	*Bubalus bubalis*	Medium	No
Goat (wild)	*Capra aegagrus*	Medium	Yes
Goat	*Capra hircus*	Medium	No
Cat	*Felis catus*	Medium	No
Giraffe (Masai)	*Giraffa tippelskirchi*	Medium	Yes
Sheep	*Ovis aries*	Medium	No
Jaguar	*Panthera onca*	Medium	Yes
Leopard	*Panthera pardus*	Medium	Yes
Tiger (Siberian)	*Panthera tigris altaica*	Medium	Yes
Cougar	*Puma concolor*	Medium	No
Cheetah	*Acinonyx jubatus*	Medium	Yes
Mole-rat (Naked)	*Heterocephalus glaber*	Medium	No
Rabbit (European)	*Oryctolagus cuniculus*	Medium	Yes
Rhinoceros (Black)	*Diceros bicornis*	Low	Yes
Panda (Giant)	*Ailuropoda melanoleuca*	Low	Yes
Camel (Bactrian)	*Camelus bactrianus*	Low	No
Camel (Dromedary)	*Camelus dromedarius*	Low	No
Bear (Grizzly)	*Ursus arctos horribilis*	Low	No
Dog	*Canis lupus familiaris*	Low	No
Bat (Lesser dawn)	*Eonycteris spelaea*	Low	No

(*Continued*)

Table 2.1 (Continued) Some notable predictions of animals susceptible to SARS-CoV-2

Animal	Latin name	Prediction of susceptibility	On IUCN red list
Donkey	Equus asinus	Low	No
Horse	Equus caballus	Low	No
Pig	Sus scrofa	Low	No
Bat (Straw-coloured fruit)	Eidolon helvum	Low	Yes
Elephant (African)	Loxodonta africana	Low	Yes
Guinea pig	Cavia tschudii	Very low	No
Bat (Lesser short-nosed fruit)	Cynopterus brachyotis	Very low	No
Pangolin (Chinese)	Manis pentadactyla	Very low	Yes
Pangolin (Sundra)	Manis javanica	Very low	Yes
Mink (European)	Mustela lutreola	Very low	Yes
Ferret	Mustela putorius furo	Very low	No
Bat (Greater horseshoe)	Rhinolophus ferrumequinum	Very low	No
Sloth (Hoffmann's two-toed)	Choloepus hoffmanni	Very low	No
Rat (Brown)	Rattus norvegicus	Very low	No
Bat (Big Brown)	Eptesicus fuscus	Very low	No
Bat (Little Brown)	Myotis lucifugus	Very low	No
Shrew (common)	Sorex araneus	Very low	No
Mink (American)	Neovison vison	Very low	No
Hedgehog (European)	Erinaceus europaeus	Very low	No
Bat (Common pipistrelle)	Pipistrellus pipistrellus	Very low	No

Data as published by Damas et al.[40]

Figure 2.7 Non-human primates, such as the Langur monkey, are likely to be susceptible to SARS-CoV-2. Photograph by Ros Rouse© (but made black and white). The original can be found at https://www.tigerloveart.co.uk.

a more holistic investigation of the molecular biology of the infection, and how the detail compares across animal species.

Of course, as mentioned, several others have also carried out similar predictions. Alexander et al.[43] looked at the ACE2 protein and identified important amino acids and did what they described as "in-depth sequence and structural analyses" and then looked at a range of species that may show susceptibility, giving each animal species a susceptibility score. They said that when looking at the data across animal species it can be concluded that SARS-COV-2 "is nearly optimal for binding ACE2 of humans." However, they do then say that:

> data suggest that cows (*Bos taurus*), Malayan pangolin (*Manis javanica*), and goats (*Capra hircus*) have intermediate susceptibility to infection, while Chinese horseshoe bats (*Rhinolophus sinicus*), horses (*Equus caballus*), and camels (*Camelus dromedarius* and *Camelus bactrianus*) have higher susceptibility.

They also point out that the only way to know for sure is to try to infect such animals and see what happens, but that this is not easy and is very costly (and of course, ethically problematic).

Structural modelling was also used by Rodrigues and colleagues,[44] and they used what was referred to as HADDOCK scores to judge susceptibility of different animal species. Birds (chicken and duck) and rodents (mouse and rat) had lower than average scores, suggesting that they are not likely to become infected. The authors point out that there are flaws in their predictions, particularly highlighting guinea pig, which was ranked extremely high and, in fact, higher than humans. But they also pointed out that their method did rank ferrets and bats high, which others had not. Dog and pangolin also scored well in this analysis. Fischhoff et al.,[45] focused, like others on ACE2, but they combined structural analysis with machine learning to look at 5,400 mammals and had a surprisingly good prediction of which species may be susceptible. In their first figure, the authors compare the predictions across 13 similar studies and then indicate how their work compares. They also correlate these data with known infections of animal species and there is a good level of prediction from their work, appearing to be much higher than others. However, there are still a couple of notable outliers, with fruit bats and rabbits not being predicted to be susceptible and yet they say that infections in these species have been confirmed, and their predictions are not aligned with those of others. It is a shame that they only concentrated on mammals here and not other vertebrates. Others also focused on ACE2. In one study, only looking at ten different animals, susceptibility was suggested for cats, cattle, and chimpanzees.[46] Lam et al.[47] focused on 26 animals that may be in close contact with humans, and then ranked their susceptibility. Zoo animals, such as non-human primates, were scored low in their analysis, meaning that they were susceptible. Interestingly, when it came to farm animals, they said that, "Of particular concern are sheep," but this has not been borne out in reality. However, the authors also say:

> our work is not aiming to provide an absolute measure of risk of infection. Rather, it should be considered an efficient method to screen a large number of animals and suggest possible susceptibility, and thereby guide further studies.

And they go on to say that more robust methods need to be employed once such screening has been carried out. Bouricha and colleagues[48] say that "all studied animals are potentially susceptible to SARS-CoV-2 infection," and a look through their methods showed that they pre-selected five species of animal that had a high sequence identity to human ACE2, including gorilla and bonobo. They then looked further down the list of homologous protein sequences and showed 16 more species with a similarity, which they say was "identity > 80% and coverage > 98%," ranging across livestock, domestic, and forest mammals, including dogs and various cats. They did not look at non-mammals, so they could not conclude anything about animals where others said there was low susceptibility, such as fish, birds, and amphibians. Interestingly, they said that a change of amino acid at position 41 of the ACE2 polypeptide, from tyrosine to histidine, reduces the ability of the SARS-CoV-2 to bind.

Qui et al.[49] did a survey of the ACE2 protein from 248 vertebrates and in their abstract said: "SARS-CoV-2 tends to utilise ACE2s of various mammals, except murines, and some birds, such as pigeon." Among their predicted susceptible animals are goats, sheep, and cattle, along with horses, cats, and mustelids. Oddly, the dog seems to have no prediction in their phylogenetic tree representation. In their further analysis, rodents were at the bottom of the susceptibility list, and low down were also reptiles and birds.

To get an idea of the likely susceptibility of animals to SARS-CoV-2, Hobbs and Reid[50] carried out a scoping review of the literature. They point out that under experimental conditions animals such as non-human primates, cats, ferrets, hamsters, and bats have been found to be able to be infected. Dogs less so, where they say that naturally dogs tend to be asymptomatic. Cats can be infected but appeared to have mild disease (it should be noted here that some big cats have died) whilst minks could die. The actual cases reported in the literature will be discussed more in depth later. Tilocca et al.[51] took a different approach and looked at the similarities in the viral protein structures. They particularly looked at epitope[52] mapping and suggested that the most closely related were ones from bat and pangolin, with avian CoV viruses being at the bottom of the similarity scale. Others too took a similar approach, with Cagliani et al.[53] concentrating on the viruses from bats and pangolins. Also concentrating on the ACE2 protein Sun and colleagues[54] did a survey of the susceptibility of 31 animals. They concluded that although the predicted binding of the virus to ACE2 was weaker in most animals than humans, some were of higher affinity than horseshoe bats. They also highlight some point mutations in the human ACE2 that could make the virus bind more tightly and therefore, one assumes, be able to infect the ACE2 containing cells more easily.

As can be seen, most of these studies focused on the ACE2 receptor on host cells as being the most important protein for prediction of infection. Two other studies will be mentioned here that also took this approach. The first is by Ma and Gong.[55] As found by others, non-human primates such as gorilla and monkey scored highly, whereas rodents (mice and rats) scored the lowest, although they only looked at nine species (gorilla, monkey, pig, cow, sheep, cat, dog, mouse, and rat). Interestingly, they say that "Cat ACE2 is predicted to have a lower S [spike] protein affinity than dog ACE2." But they go on to say that they think all these animals are likely to be susceptible, little or much. In the second study by Sang et al.[56] they say that they have refined their predictions to "fit recent experimental validation" (accepted for publication August 2020). They conclude that domestic animals, such as dogs, pigs, cattle, and goats, are unlikely to be a problem in amplifying the virus as the pandemic unfolds, unless mutations of some form alter this prediction.

It can be seen, therefore, that there was a variety of approaches taken here, but most of the studies concentrated on investigating the similarities in the ACE2 protein across species, as compared to that seen in humans. There was some consensus of opinion across these papers, with non-human primates always scoring high, and some vertebrates, such as reptiles and birds, scoring low. However, animals scoring in the middle of the range could not be easily ranked, and often they were in different orders. It was hard to have a clear take-home message here. In Chapter 3, we will concentrate on the real world, and bearing in mind what the molecular biologists predicted is interesting in hindsight. Perhaps, having such a range of studies and by learning from these, we can improve on such predictions in the future.

2.5 What data are missing?

Estimates vary widely, but one set of data suggests there are approximately 2.16 million species of animals on the planet, just over 3% of which are vertebrates, and only 6,596 of which are mammals.[57] Above researchers used only a handful of these for further genetic analysis, so we are left to assume that if we use two fish species, we can extrapolate that to all fish species. This is a big stretch. We, like all others, cherry-picked which species to look at. There are obvious ones. Gorilla is likely to be the same as us. Dogs and cats (domestic) live with us. Minks are known to be a problem (see below). But how do we select the others? Even the Damas et al. paper looking at 410 species only scratched the surface of the problem.

So why don't we look at more? First, it is a lot of work and of limited use, pulling in sequences and aligning them, although it is no doubt able to be automated to a certain extent. Having said that, alignment tools are limited to how many sequences can be used in any one run, at least at the moment. As of writing this, *Clustal Omega* states that it is limited to "4000 sequences or a maximum file size of 4 MB." Second, not all the data needed are available. Getting the protein sequence before we start this sort of analysis requires either cloning of the DNA/RNA sequences encoding the protein or sequencing the protein itself. None of these are trivial, although the nucleotide database is growing at an astonishing rate, especially as we now have Next Generation Sequencing (NGS) technology. However, much of this data relies on researchers working on individual proteins and genes rather than whole genomes, so even if some of the protein sequences are available for a species of animal, it does not guarantee that the proteins you wish to study are there, such as ACE2.

Some species are rare. Some are hard to work with, and there are ethical issues about working on endangered species. No one wishes to harm or kill something on the International Union for Conservation of Nature (IUCN) Red List.[58] It is likely, therefore, that required data will not become available for many animal species, unless samples are taken during veterinary inspections, for example.

Therefore, it is likely that it will be some time before all species that are likely to be susceptible to viruses such as SARS-CoV-2 are available to be able to carry out quick bioinformatic analysis.

2.6 How can analysis using *in silico* techniques be improved to aid in future predictions?

Much of the work in the bioinformatic approach above is simple sequence alignments, as described above. Within those sequences, the amino acids that are most important in the functions bestowed by the protein need to be identified, and this is not always easy to do. Each amino acid has its own unique characteristics, determined by its side chain. However, amino acids can be grouped by their characteristics.[59] Therefore, conserved substitutions need to be considered – some researchers do this, others don't (see molecular biology section above).

In the same vein, the structures of the proteins need to be more commonly considered. Again, some researchers do, and others do not. Without knowing the protein structures, it is hard to predict if proteins can fit together, i.e., interact. For some of the proteins involved in the viral life cycle, the structures may not be yet known, and therefore this can cause an issue if *in silico* methods are going to be useful for prediction; however, the rise of artificial intelligence (AI)-driven structure prediction software such as *Alphafold*[60] is a solution to this.

The gene sequences encoding for proteins contain polymorphisms. That means that even if two individuals have the same gene, e.g., two of the people authoring this text, it does not mean that the gene sequences are the same. There may be small changes along the gene sequence. This is often at one point, i.e., at one base along the DNA strand, and such polymorphisms are referred to as single nucleotide polymorphisms, or SNPs (pronounced snips). Sometimes, because the genetic code allows for redundancy,[61] there can be a change in the DNA sequence but no change in the protein. However, at other times a different amino acid may be used in the protein. This again may have little consequence if the amino acid is in a non-functioning region. However, if the amino acid is part of an active site, or a binding region, alteration of the protein's function might result. A protein may become less able to fulfil its role, but, on the other hand, a protein might be better than the original. In COVID-19, some polymorphisms, such as in the GSTO proteins discussed above, seem to increase the susceptibility of the individual carrying that polymorphism to the virus, or the symptoms are worse. To fully understand if animals are susceptible, such polymorphisms ought to be taken into account. However, this is almost an impossible task, as the sequences of all the relevant proteins in that individual animal would need to be known. On a simple level, known polymorphisms in humans may be able to be searched in animal sequences from the database to perhaps get an indication of susceptibility. This is the approach we took in our paper.[62] However, it is a rather crude and naïve analysis, and it would take a considerably greater amount of work, both in the laboratory and *in silico*, to make robust predictions of the susceptibility of a chosen animal to SARS-CoV-2. Known polymorphisms can be found on a public database.[63]

Finally, there needs to be more of a holistic approach. All the proteins involved need to be considered. Here, a few are discussed, but by no means all of them. The virus has to invade and hijack the cell, and numerous proteins will be involved. We have considered mainly interacting proteins allowing viral entry, but other proteins are involved in the viral replication mechanisms, such as the nsp14 exoribonuclease.[64] Let's think about a simple thought experiment. A virus is not predicted, using the bioinformatics, to be very likely to invade the cell of a snail, but the proteins involved in the replication of the virus once it is inside the cell are super-efficient at making viral copies. A very low amount of virus entering the snail's cells may lead to a sustainable production of future viruses. Should we be concerned that the snail can be infected and spread the disease? (Please note, in reality, snails are very unlikely to be SARS-Cov-2 positive). It is the whole system that needs to be considered. What is needed is modelling of the whole viral life cycle, from arriving at the cell surface to being released as copies, and including all the players in between. Are the proteins in the snail likely to partake in this activity? Is there something in the snail that makes it impossible? Or is there something in the snail that can compensate for the poor efficiency of other activities, which means that, when taken as a whole, the snail will be infected and allow the virus to be propagated. If we had such holistic modelling then we may be in a position to predict in the future whether viruses such as SARS-CoV-2 would be a problem in any chosen animal species. This is a big ask,[65] but researchers are seeking to create a working cell *in silico*,[66,67] so this may not be as far-fetched as one might think.

2.7 Conclusions

There is much known about the interaction of the virus with its host cells. This interaction is dictated by the presence of proteins "sticking out" of the virus envelope (the spike proteins) and the presence of proteins on the surface of the host cells with which the viral spike proteins can interact. One of the host proteins thought to be most influential here is ACE2, but this is only one of many, and viral entry will be a well-orchestrated mechanism involving a range of proteins. Once inside the cell, the virus will need to hijack the cell to make more copies of itself, and therefore, many more proteins will be involved.

The SARS-CoV-2 virus seems well suited to interact with and invade human cells, but what about other animals?

Concentrating on the ACE2 protein as a potential marker of predicting an animal's susceptibility to the virus has had limited use. Some animals, perhaps the obvious ones such as non-human primates, can be predicted to be susceptible. Others, such as amphibians, fish, birds, and reptiles, appear to be predicted to be safe from the virus. However, for a whole range of animals on the evolutionary scale between primates and birds, there is little prediction that seems to be useful. In the next chapter, when the real world is discussed, it will be found that the predictions were not necessarily accurate.

The biggest issue here is that we do not know much of the information needed to make accurate predictions. For the vast majority of animal species, the DNA[68] or protein sequences are not available, and therefore there is no opportunity to use bioinformatics for making predictions. Furthermore, much of the work using bioinformatics uses simple multiple alignments of sequences, and this is of limited value. It is the folding of the proteins, as well as their amino acid sequence (which of course also dictates the folding), that bestows on the protein its function, including how it may interact with other proteins. Without such folding analysis, the alignments are not very useful, but carrying out the structural analysis of all the proteins needed is not easy, takes a long time (relatively), and is more expensive. Perhaps by combining multiple forms of sequence data with AI-driven structural predictions and search algorithms, these problems can be overcome.

Lastly, it is not clear which proteins to target for these predictions. ACE2 became an obvious choice early in the pandemic. But ACE2 does not function alone, and proteins such as furin were given more prominence. But who would have instantly guessed that GSTO, being an enzyme involved in oxidative stress metabolism, should be considered?

This is not an exhaustive list of uses for predictive bioinformatics, but the discussion does highlight both the potential and the flaws in this approach. In the next chapter, we will discuss what happened in the real world.

Notes

1 Viruses are not deemed to be truly living so the phrase "life cycle" is being used loosely here.
2 For an excellent fictional work on such warfare see *Victus; The Fall of Barcelona, a Novel*, by Albert Sanchez Pinol. ISBN: 0062323962.

3 Gallo Marin, B., Aghagoli, G., Lavine, K., Yang, L., Siff, E.J., Chiang, S.S., Salazar-Mather, T.P., Dumenco, L., Savaria, M.C., Aung, S.N. and Flanigan, T. (2021) Predictors of COVID-19 severity: A literature review. *Reviews in Medical Virology, 31*, 1–10.

4 LibreTexts, biology: 1.4: Crash course in molecular biology: https://bio.libretexts.org/Bookshelves/Computational_Biology/Book%3A_Computational_Biology_-_Genomes_Networks_and_Evolution_(Kellis_et_al.)/01%3A_Introduction_to_the_Course/1.04%3A_Crash_Course_in_Molecular_Biology (Accessed 13/04/23).

5 It is worth noting here that sulfur is now recognised as being spelt in the "American" way, with an "f," not a "ph."

6 NIH: Nucleotide, definition: https://www.genome.gov/genetics-glossary/Nucleotide#:~:text=A%20nucleotide%20is%20the%20basic,)%20and%20thymine%20(T (Accessed 26/05/23).

7 Watson, J.D. and Crick, F.H.C. (1953) A structure for deoxyribose nucleic acid. *Nature, 171*, 737–738.

8 Britannica: Rosalind Franklin: https://www.britannica.com/biography/Rosalind-Franklin (Accessed 13/04/23).

9 AKJournal: Scientometric portrait of nobel laureate Dorothy Crowfoot Hodgkin: https://akjournals.com/view/journals/11192/45/2/article-p233.xml (Accessed 13/04/23).

10 Salzberg, S.L. (2018) Open questions: How many genes do we have? *BMC Biology, 16*, 1–3.

11 Ponomarenko, E.A., Poverennaya, E.V., Ilgisonis, E.V., Pyatnitskiy, M.A., Kopylov, A.T., Zgoda, V.G., Lisitsa, A.V. and Archakov, A.I. (2016) The size of the human proteome: The width and depth. *International Journal of Analytical Chemistry.* https://doi.org/10.1155/2016/7436849.

12 Kim, D., Lee, J.Y., Yang, J.S., Kim, J.W., Kim, V.N. and Chang, H. (2020) The architecture of SARS-CoV-2 transcriptome. *Cell, 181*, 914–921.

13 Huang, Y., Yang, C., Xu, X.F., Xu, W. and Liu, S.W. (2020) Structural and functional properties of SARS-CoV-2 spike protein: Potential antivirus drug development for COVID-19. *Acta Pharmacologica Sinica, 41*, 1141–1149.

14 For full details of this protein structure see: https://www.rcsb.org/structure/7U0N (Accessed 03/11/22).

15 Shang, J., Ye, G., Shi, K., Wan, Y., Luo, C., Aihara, H., Geng, Q., Auerbach, A. and Li, F. (2020) Structural basis of receptor recognition by SARS-CoV-2. *Nature, 581*, 221–224. https://doi.org/10.1038/s41586-020-2179-y.

16 The genome of an organism is the full complement of genetic material it contains, which is capable of encoding a new individual of that species.

17 There are several multiple sequence alignment tools, but a popular one is *Clustal Omega*: https://www.ebi.ac.uk/Tools/msa/clustalo/ (Accessed 04/11/22).

18 Sun, J., He, W.-T., Wang, L., Lai, A., Ji, X., Zhai, X., Li, G., Suchard, M.A., Tian, J., Zhou, J. Veit, M. and Su, S. (2020) COVID-19: Epidemiology, evolution, and cross-disciplinary perspectives. *Trends in Molecular Medicine, 26*, 483–495. https://doi.org/10.1016/j.molmed.2020.02.008.

19 Damas, J., Hughes, G.M., Keough, K.C., Painter, C.A., Persky, N.S., Corbo, M., Hiller, M., Koepfli, K.P., Pfenning, A.R., Zhao, H. and Genereux, D.P. (2020) Broad host range of SARS-CoV-2 predicted by comparative and structural analysis of ACE2 in vertebrates. *Proceedings of the National Academy of Sciences, 117*, 22311–22322.

20 Wan, Y., Shang, J., Graham, R., Baric, R.S. and Li, F. (2020) Receptor recognition by the novel coronavirus from Wuhan: An analysis based on decade-long structural studies of SARS coronavirus. *Journal of Virology*, *94*, e00127–20.

21 This figure was first published in July 2020 in an *Animal Welfare Research Network* article and can be found here: https://awrn.co.uk/2020/07/29/predicting-the-susceptibility-of-animals-to-covid-19/ (Accessed 03/11/22).

22 Kumar, A., Pandey, S.N., Pareek, V., Narayan, R.K., Faiq, M.A. and Kumari, C. (2021) Predicting susceptibility for SARS-CoV-2 infection in domestic and wildlife animals using ACE2 protein sequence homology. *Zoo Biology*, *40*, 79–85.

23 Hancock, J.T., Rouse, R.C., Stone, E. and Greenhough, A. (2021) Interacting proteins, polymorphisms and the susceptibility of animals to SARS-CoV-2. *Animals*, *11*, 797.

24 Hoffmann, M., Kleine-Weber, H., Schroeder, S., Krüger, N., Herrler, T., Erichsen, S., Schiergens, T.S., Herrler, G., Wu, N.H., Nitsche, A. and Müller, M.A. (2020) SARS-CoV-2 cell entry depends on ACE2 and TMPRSS2 and is blocked by a clinically proven protease inhibitor. *Cell*, *181*, 271–280.

25 NCBI is the *National Center for Biotechnology Information* and can been found here: https://www.ncbi.nlm.nih.gov/ (Accessed 04/11/22).

26 Wu, C., Zheng, M., Yang, Y., Gu, X., Yang, K., Li, M., Liu, Y., Zhang, Q., Zhang, P., Wang, Y. et al. (2022) Furin: A potential therapeutic target for COVID-19. *iScience*, *23*, 101642.

27 Thomas, G. (2002) Furin at the cutting edge: From protein traffic to embryogenesis and disease. *Nature Reviews Molecular Cell Biology*, *3*, 753–766.

28 Daly, J.L., Simonetti, B., Klein, K., Chen, K.E., Williamson, M.K., Antón-Plágaro, C., Shoemark, D.K., Simón-Gracia, L., Bauer, M., Hollandi, R. and Greber, U.F. (2020) Neuropilin-1 is a host factor for SARS-CoV-2 infection. *Science*, *370*, 861–865.

29 Polymorphisms can be thought of as changes in the genetic sequence that can be lived with, in that they are not usually life threatening and therefore can be passed on through reproduction. We all have them.

30 Senapati, S., Banerjee, P., Bhagavatula, S., Kushwaha, P.P. and Kumar, S. (2021) Contributions of human ACE2 and TMPRSS2 in determining host–pathogen interaction of COVID-19. *Journal of Genetics*, *100*, 1–16.

31 Da Cunha, J.P.C., Galante, P.A.F., De Souza, J.E., De Souza, R.F., Carvalho, P.M., Ohara, D.T., Moura, R.P., Oba-Shinja, S.M., Marie, S.K.N., Silva Jr, W.A. and Perez, R.O. (2009) Bioinformatics construction of the human cell surfaceome. *Proceedings of the National Academy of Sciences*, *106*, 16752–16757.

32 Forman, H.J. and Zhang, H. (2021) Targeting oxidative stress in disease: Promise and limitations of antioxidant therapy. *Nature Reviews Drug Discovery*, *20*, 689–709.

33 Guan, R., Van Le, Q., Yang, H., Zhang, D., Gu, H., Yang, Y., Sonne, C., Lam, S.S., Zhong, J., Jianguang, Z. and Liu, R. (2021) A review of dietary phytochemicals and their relation to oxidative stress and human diseases. *Chemosphere*, *271*, 129499.

34 A disulfide bond is a sulfur-to-sulfur bond, which gives stability to a protein as it creates a strong bridge across two parts of a protein.

35 Some interesting facts about ATP, from Lane, N. (2015) *The Vital Question: Energy, Evolution, and the Origins of Complex Life*. W.W. Norton & Company. A single cell consumes around 10 million molecules of ATP every second. The turnover of ATP is 60-100 kg per day, so about your own body weight.

We only have about 60 g of ATP, so each ATP molecule is recharged once or twice every minute.

36 Djukic, T., Stevanovic, G., Coric, V., Bukumiric, Z., Pljesa-Ercegovac, M., Matic, M., Jerotic, D., Todorovic, N., Asanin, M., Ercegovac, M. and Ranin, J. (2022) GSTO1, GSTO2 and ACE2 polymorphisms modify susceptibility to developing COVID-19. *Journal of Personalized Medicine, 12,* 458.

37 Molecular biologists often talk about downstream and upstream. Downstream basically means following, and upstream, before. These terms are often used when discussing DNA/RNA/protein sequences or cell signalling events.

38 Hancock, J.T., Veal, D., Craig, T.J. and Rouse, R.C. (2022) Do SNPs in glutathione S-transferase-Omega allow predictions of the susceptibility of vertebrates to SARS-CoV-2?. *Reactive Oxygen Species, 12,* c14–c29.

39 This number refers to the entry in the SNP database, and can be used for either searching or comparative analysis.

40 Damas, J., Hughes, G.M., Keough, K.C., Painter, C.A., Persky, N.S., Corbo, M. et al. (2020) Broad host range of SARS-CoV-2 predicted by comparative and structural analysis of ACE2 in vertebrates. *PNAS, 117,* 22311–22322.

41 Zhou, P., Yang, X.-L., Wang, X.-G., Hu, B., Zhang, L., Zhang, W., Si, H.-R., Zhu, Y., Li, B., Huang, C.-L. et al. (2020) A pneumonia outbreak associated with a new coronavirus of probable bat origin. *Nature, 579,* 270–273.

42 WolrdAtlas, How many species of bat are there? https://www.worldatlas.com/articles/bat-species.html (Accessed 19/12/22).

43 Alexander, M.R., Schoeder, C.T., Brown, J.A., Smart, C.D., Moth, C., Wikswo, J.P., Capra, J.A., Meiler, J., Chen, W. and Madhur, M.S. (2020) Which animals are at risk? Predicting species susceptibility to Covid-19. *bioRxiv*.

44 Rodrigues, J.P., Barrera-Vilarmau, S., Mc Teixeira, J., Sorokina, M., Seckel, E., Kastritis, P.L. and Levitt, M. (2020) Insights on cross-species transmission of SARS-CoV-2 from structural modeling. *PLoS Computational Biology, 16,* e1008449.

45 Fischhoff, I.R., Castellanos, A.A., Rodrigues, J.P., Varsani, A. and Han, B.A. (2021) Predicting the zoonotic capacity of mammal species for SARS-CoV-2. *bioRxiv.* This paper was later as: Fischhoff, I.R., Castellanos, A.A., Rodrigues, J.P., Varsani, A. and Han, B.A. (2021) Predicting the zoonotic capacity of mammals to transmit SARS-CoV-2. *Proceedings of the Royal Society B, 288,* 20211651.

46 Shen, M., Liu, C., Xu, R., Ruan, Z., Zhao, S., Zhang, H., Wang, W., Huang, X., Yang, L., Tang, Y. and Yang, T. (2020) Predicting the animal susceptibility and therapeutic drugs to SARS-CoV-2 based on spike glycoprotein combined with ACE2. *Frontiers in Genetics, 11,* 575012.

47 Lam, S.D., Bordin, N., Waman, V.P., Scholes, H.M., Ashford, P., Sen, N., Van Dorp, L., Rauer, C., Dawson, N.L., Pang, C.S.M. and Abbasian, M. (2020) SARS-CoV-2 spike protein predicted to form complexes with host receptor protein orthologues from a broad range of mammals. *Scientific Reports, 10,* 1–14.

48 Bouricha, E.M., Hakmi, M., Akachar, J., Belyamani, L. and Ibrahimi, A. (2020) *In silico* analysis of ACE2 orthologues to predict animal host range with high susceptibility to SARS-CoV-2. *3 Biotech, 10,* 1–8.

49 Qiu, Y., Zhao, Y.B., Wang, Q., Li, J.Y., Zhou, Z.J., Liao, C.H. and Ge, X.Y. (2020) Predicting the angiotensin converting enzyme 2 (ACE2) utilizing capability as the receptor of SARS-CoV-2. *Microbes and Infection, 22,* 221–225.

50 Hobbs, E.C. and Reid, T.J. (2021) Animals and SARS-CoV-2: Species susceptibility and viral transmission in experimental and natural conditions, and

the potential implications for community transmission. *Transboundary and Emerging Diseases, 68*, 1850–1867.

51 Tilocca, B., Soggiu, A., Sanguinetti, M., Babini, G., De Maio, F., Britti, D., Zecconi, A., Bonizzi, L., Urbani, A. and Roncada, P. (2020) Immunoinformatic analysis of the SARS-CoV-2 envelope protein as a strategy to assess cross-protection against COVID-19. *Microbes and Infection, 22*, 182–187.

52 A region of a protein to which antibodies are most likely to be raised in an immune response.

53 Cagliani, R., Forni, D., Clerici, M. and Sironi, M. (2020) Coding potential and sequence conservation of SARS-CoV-2 and related animal viruses. *Infection, Genetics and Evolution, 83*, 104353.

54 Sun, H., Wang, A., Wang, L., Wang, B., Tian, G., Yang, J. and Liao, M. (2022) Systematic tracing of susceptible animals to SARS-CoV-2 by a bioinformatics framework. *Frontiers in Microbiology, 13*, p.781770.

55 Ma, C. and Gong, C. (2021) ACE2 models of frequently contacted animals provide clues of their SARS-CoV-2 S protein affinity and viral susceptibility. *Journal of Medical Virology, 93*, 4469–4479.

56 Sang, E.R., Tian, Y., Gong, Y., Miller, L.C. and Sang, Y. (2020) Integrate structural analysis, isoform diversity, and interferon-inductive propensity of ACE2 to predict SARS-CoV2 susceptibility in vertebrates. *Heliyon, 6*, e04818.

57 Hannah Ritchie (2022) - "How many species are there?". Published online at OurWorldInData.org. Retrieved from: https://ourworldindata.org/how-many-species-are-there (Accessed 14/08/23).

58 The IUCN Red List: https://www.iucnredlist.org/ (Accessed 04/11/22).

59 Stanfel, L.E. (1996) A new approach to clustering the amino acid. *Journal of Theoretical Biology, 183*, 195–205.

60 AlphaFold Protein Structure Database: https://alphafold.com/ (Accessed 12/04/23).

61 Redundancy: Amino acids are encoded for by three bases, called codons. Often more than one codon will code for the same amino acid, so some changes in the DNA sequence may alter the codon sequence but it may still encode for the same amino acid, and so no change in the protein will be seen. The amino acid serine, e.g., has six codons. A copy of the genetic code can be found here: https://www.thoughtco.com/genetic-code-373449 (Accessed 15/12/22).

62 Hancock, J.T., Rouse, R.C., Stone, E. and Greenhough, A. (2021) Interacting proteins, polymorphisms and the susceptibility of animals to SARS-CoV-2. *Animals, 11*, 797.

63 Polymorphism database: http://www.ncbi.nlm.nih.gov/SNP (Accessed 19/12/22). More information about this data can be found at: Smigielski, E.M., Sirotkin, K., Ward, M. and Sherry, S.T. (2000) dbSNP: A database of single nucleotide polymorphisms. *Nucleic Acids Research, 28*, 352–355.

64 Ogando, N.S., Zevenhoven-Dobbe, J.C., van der Meer, Y., Bredenbeek, P.J., Posthuma, C.C. and Snijder, E.J. (2020) The enzymatic activity of the nsp14 exoribonuclease is critical for replication of MERS-CoV and SARS-CoV-2. *Journal of Virology, 94*, e01246–20.

65 Palsson, B. (2000) The challenges of in silico biology. *Nature Biotechnology, 18*, 1147–1150.

66 Normile, D. (1999) Building working cells 'in silico'. *Science, 284*, 80–81.

67 Wang, C., Li, S., Ademiloye, A.S. and Nithiarasu, P. (2021) Biomechanics of cells and subcellular components: A comprehensive review of computational

models and applications. *International Journal for Numerical Methods in Biomedical Engineering, 37,* e3520.

68 The DNA sequence can easily be translated into the amino acid sequence using the genetic code, and can be quickly done using computer programmes.

3

Animals that were infected in the real world

3.1 Introduction

In the last chapter, the use of *in silico* technologies was discussed as a potential way to predict if animals can be infected with the SARS-CoV-2 virus. Predictions are all very well, but what is needed is knowledge of real-world events. As more and more humans became infected, were the animals really at risk or were they safe?

The UK prime minister at the time of the pandemic, Boris Johnson, was reported as saying that he was worried about his dog, and even asked if it could be tested for COVID-19.[1] Was he worrying unnecessarily?

It may have been that Mr Johnson's dog was infected with a coronavirus, as there are canine-specific variants of these viruses. Canine respiratory coronavirus (CRCoV) is a betacoronavirus, which is distinct from canine enteric coronavirus, an alphacoronavirus.[2] CRCoV is often associated with what is known as "kennel cough,"[3] causing inflammation and damage to the respiratory epithelium, and so a dog is likely to be symptomatic.

This highlights two issues. First, just because an animal is showing what may appear to be COVID-19 symptoms does not mean it is infected with SARS-CoV-2. Even during the pandemic humans had cold and influenza; these other diseases did not simply disappear. All animals, including humans, are bombarded by pathogens all the time, and SARS-CoV-2 is just one of them. Second, the fact that Mr Johnson had to ask for his dog to be tested shows that animal testing was not common. Interestingly, when India announced the launch of its animal COVID-19 vaccine, it also announced a testing kit for dogs called the CAN-CoV-2

DOI: 10.1201/9781003427254-3

ELISA Kit.[4] Even so, even now no one can be sure which animals and how many were infected with SARS-CoV-2. However, what is known is that many did become infected, and several died from that infection. Here we will have a look at some of the evidence that animals were found to be infected, and in later chapters we will discuss some of the indirect consequences of this pandemic.

3.2 Animals that were infected

There were several groups of animals that were of concern during the pandemic: companion animals, those in captivity (including those being farmed), and those in the wild. Although some of what is known in some species might overlap these groupings, the groups will be considered separately below.

3.2.1 Companion animals found to be infected with SARS-Cov-2

Companion animals, by definition, will be in close contact with their humans, who are likely petting and cuddling them as well as handling their food, and even breathing over it, or sharing their own food. Therefore, if such animals are susceptible, then there is a reasonable likelihood of infection. The first examples of companion animals becoming infected in the USA were cats in New York, for example, reported on 22 April 2020.[5] There were some worries about pets increasing the spread of SARS-CoV-2, with the UK Government even considering a cull of cats.[6] Such potential policies showed that there was a possible willingness to sacrifice animals to protect human health and how little was known for sure in the early phases of the pandemic.

Dogs have been shown to be infected,[7] although the symptoms seen in dogs tends to be mild. There have been some reports that suggest that in some dogs and cats, infection with SARS-CoV-2 is linked to inflammation of heart tissue, i.e., myocarditis.[8,9] Such a link was first suggested by Ferasin and colleagues,[10] but it has yet to be fully understood. Interestingly, there has been a suggested link in humans between COVID-19 and endocarditis,[11] an infection of the inner lining of the heart, usually caused by a bacterium.

There appears to be no evidence that dogs can infect each other or humans. Having said that, in February 2023, Kamel et al., in a review of the interactions between dogs and humans during the pandemic, stated that "risk of transmission by domestic dogs remains a concern."[12]

However, there is also no evidence that dogs cannot act as a surface from which the virus can be picked up and therefore infect humans who are petting them. Anecdotally there has been some worry that dogs may carry the virus between people either in the same household or between homes, for example in a block of flats (apartments). It always seemed odd during times of COVID control to see humans socially distancing in parks whilst their dogs mingled with both other dogs and all the humans, as though they were immune to the virus. There was no testing of dogs, either from their respiratory tracts or from their fur, so the consequence of the mingling of dogs between households will remain unknown.

On the other hand, cats can become infected[13] and pass the virus between individuals. There is no evidence that cats can pass the virus back to humans, but the same arguments as given for dogs above may still apply. Symptoms in cats are more severe, and, as discussed for dogs, there was some evidence of a link to heart problems in cats. Some felines have died, as discussed for big cats below, although pet cats were also involved. For example, in the United Kingdom the first report of a cat becoming infected from human contact and then dying was April 2021. The four-month-old cat was found to have SARS-CoV-2 in its lungs. A newspaper at the time was encouraging people to stop cuddling their pets,[14] although it is doubtful whether such articles would have had any impact.

However, also in a manner similar to dogs, there are feline-specific coronaviruses (FCoV),[15] and just because a cat seems to have symptoms does not automatically mean that it is infected with SARS-CoV-2. Carpenter and colleagues suggested how to determine the role of SARS-CoV-2 in deaths in cats and dogs, using ten case studies, concluding that their work "provides evidence that SARS-CoV-2 can, in rare circumstances, cause or contribute to death in pets."[16]

One animal, which came as a bit of a surprise, that was reported as being SARS-CoV-2 positive was the hamster,[17,18] although experimental studies have shown that Golden Syrian hamsters (*Mesocricetus auratus*) have a high susceptibility, and the authors were concerned that hamsters may be a potential route for the virus to infect humans.[19] In the real-world, Syrian hamsters were found to be positive in a pet shop and in a warehouse. They were infected with what was thought to be the Delta variant of the virus. Interestingly, numerous dwarf hamsters, rabbits, guinea pigs, chinchilla, and mice were also tested, and none were found to be positive. The worry here was that the people handling the hamsters were subsequently found to be COVID-19 positive, and that hamsters had transmitted the Delta variant to humans. About 50 people were affected. As a result, the local authorities in Hong Kong decided to have approximately 2,000 hamsters killed to stop the spread of the virus via this route.

3.2.2 Animals found to be infected with SARS-CoV-2 in captivity

Animals kept in captivity not only are usually in relatively close contact with humans, but also even during a pandemic there was a need to maintain human interactions to be fed, watered, and cleaned. Therefore, animals in zoos and farms still had the risk of being infected if they were susceptible. Other worries include the species being kept. Many zoos keep non-human primates – for example, our own local Bristol Zoological Society is famous for its gorilla colony.[20] As discussed above, the genomes of these animals are most likely to be similar to humans, and therefore their proteins, such as the ACE2 protein, are most likely to be similar, leading to a conclusion that such animals are going to be susceptible to the virus. Furthermore, often animals in captivity are sadly kept in close proximity to each other, as can be seen in Figure 3.1. Therefore, once one individual is infected, the virus can easy spread, either from airborne droplets or from surfaces.

So how did the captivity of animals impact upon those individual animals? Considering that animal keepers may have been infected, and therefore able to infect the animals in their charge, and also considering that many animals in captivity are endangered, for example in zoos, was there an impact of SARS-CoV-2 that we should be concerned about? And if so, what can we do better for any future pandemic?

Gorillas are likely to be susceptible, as shown in Table 2.1, and they are indeed able to be infected with SARS-CoV-2. Thirteen Western lowland gorillas were tested positive at Atlanta Zoo. It was thought that the virus was

Figure 3.1 Minks in cages in a mink farm. Photograph from Shutterstock, curtesy of Nicolai Dybdal, but has been made black and white.

transmitted from a vaccinated zoo keeper who was wearing appropriate PPE. However, as the news article[21] said, the gorillas at the zoo "live in close proximity to each other in four groups," and therefore they could not easily be isolated and the virus could easily spread across the colony. In San Diego Zoo there were reports of gorillas testing positive, with *National Geographic* saying that these were the "first non-human primates with confirmed cases."[22] During 2021, Casal and Singer highlighted a stark warning: "If the current pandemic continues, it could give several species of great apes the final push into extinction."[23] They point out that countries in which apes live cannot be expected to finance the conservation efforts needed, and that other countries need to step in to help (Figure 3.2).

Figure 3.2 Lowland Gorilla. These animals are likely to be susceptible to SARS-CoV-2. Painting by Ros Rouse© (but made black and white). The original can be found at https://www.tigerloveart.co.uk/.

As it is known that cats can become infected with SARS-CoV-2, it came as no surprise that several big cats kept in captivity were found to have tested positive. Four lions tested positive at Barcelona Zoo, for example.[24,25] Three were females and one a male, and they were tested using PCR. Only mild symptoms were reported for these animals. Lions and a puma in a private zoo in Johannesburg, South Africa, were reported to have been sick for about three weeks. Although for most animals the symptoms were mild, one lion developed pneumonia, and it took seven weeks for the viral tests to become negative. The animals were probably infected by their handlers.[26,27]

At the Smithsonian National Zoo in Washington six African lions, a Sumatran tiger, and two Amur tigers were found to be COVID-19 positive in preliminary tests on their faeces. The animals were reported to be "lethargic and were observed coughing, sneezing and not finishing meals."[28] Eight lions were also reported to be SARS-Cov-2 positive in India, at Nehru Zoological Park in Hyderabad.[29] Two lions at the Lion Safari Park, Etawah, Uttar Pradesh, India, and one at Nahargarh Biological Park, India, were found to be positive with the B.1.617.2 (Delta) variant of SARS-CoV-2.[30] Also in India, at Arignar Anna Zoological Park in Chennai, Asian lions (*Panthera leo persica*) in a safari park showed symptoms and then were found to be positive for the virus, also with the B.1.617.2 (Delta) variant being found to be involved.[31] A lion was also reported to be viral positive in Sri Lanka.[32] A cougar and three tigers also tested positive at the In-Sync Exotics Wildlife Rescue and Educational Center in Texas.[33] In New York, at the Bronx Zoo, four tigers and three lions showed COVID-19-like symptoms and there was evidence of SARS-CoV-2 in the faeces and respiratory secretions from all the animals.[34] A separate report from the same zoo said that two Malayan tigers (*Panthera tigris jacksoni*), two Amur tigers (*Panthera tigris altaica*), and three lions (*Panthera leo krugeri*) were infected.[35] Clearly these were not rare outbreaks with lions and other big cats being found to be infected in many places in the world. A look back at Table 2.1 shows that lions were only in the medium susceptibility category as predicted by Damas et al., casting doubt on the robustness of such predictions (Figure 3.3).

The situation for felines was highlighted when deaths of animals were reported. For example, in Valandur Zoo in India, two deaths of lions from SARS-CoV-2 were reported.[36] Three snow leopards (*Panthera uncia*) were said to have died from complications following a COVID-19 infection at Nebraska Zoo.[37]

One of the biggest issues with animals becoming positive with the SARS-CoV-2 virus was with mink. Minks are farmed for their fur, and, as such, they are often held in tight and crowded conditions, as seen in Figure 3.1. Therefore, if one animal becomes infected, a spread of the

Figure 3.3 Big cats, such as the tiger, were susceptible to SARS-CoV-2. Painting by Ros Rouse©, although made black and white. The original can be found at https://www.tigerloveart.co.uk.

virus is likely. Concerns first arose amid reports that the minks could infect humans who were working at these farms,[38] and there seems to be good scientific evidence for this.[39] Furthermore, mink may be infected in a sub-clinical (asymptomatic) manner, meaning that the virus can be transmitted between animals undetected. And if mink can be infected, does this mean other mustelids (such as ferrets) can also be susceptible to the virus?

According to Mike Brown of the International Fur Federation, USA, over 12,000 out of approximately three million farmed mink in the country have died from COVID-19 before they were slaughtered for their fur.[40] This is an astonishing number, especially when you look back at Table 2.1 and see that minks (and ferrets) were thought to be very low on the susceptibility scale.

Many countries were affected by SARS-CoV-2 outbreaks at mink farms, including the Netherlands (which has now permanently banned mink farming, on an accelerated schedule brought about by the pandemic, bringing the planned ban forward from 2024[41]), Denmark, United States, Spain, Canada, Lithuania, France, Greece, Italy, Poland,[42] and Sweden. As a result of virus outbreaks at farms, culling programmes were implemented. In the Netherlands in April and May 2020 two farms were reported to have infected mink,[43] and with viral-derived RNA in the dust at the farms the authors of the report were concerned that there may be an issue for human health. This was also reported by Lu et al.,[44] who said that the movement of people and the distance between farms were contributing factors in the spread of the virus. Pomorska-Mól et al. suggest that:

> minks could represent potentially dangerous, not always recognized, animal reservoir for SARS-CoV-2.[45]

The situation of SARS-CoV-2 spreading through mink farms has also been discussed by others. Fortunately, although there was a worry about human health being threatened in farms there was little evidence that the infection potential spread outside the farms to the local environment and local communities.[46] On the other hand, dogs and feral cats at mink farms were found to be infected, and it was concluded that the feral cats (as opposed to the companion cats that were also tested) were infected by contact with mink. The authors estimated that feral cats had a 12% chance of becoming infected based on a mean 20-day exposure time and assuming no cat-to-cat transmission.[47] Sharun and colleagues[48] suggest some mitigating actions that could be taken to improve the risk for humans in the farms, such as better surveillance and monitoring, and also a call for a more One Health approach, which we will revisit later.

One of the worries about the spread of the virus in mink farms was that it may propagate a new variant. One such genetic variant of the virus, dubbed the "cluster five," was detected in mink farms in Denmark in November 2020. Indeed, this variant was subsequently detected in 12 human cases of the virus,[49] although whether this was direct transmission from mink is unknown. This variant is of particular concern as it contained a mutation (Y453F) in the receptor binding domain of the spike protein.[50] This enables the virus to bind to the ACE2 protein on the host cells with four times the affinity than the original virus, meaning that it might have the potential to spread and be transmitted more. Coupled with evidence that people working on mink farms had a higher risk of SARS-CoV-2 infection than the general population, this and other evidence led to the nationwide cull of all farmed mink in Denmark.

As discussed in Section 6.3, the reporting of SARS-CoV-2 outbreaks in mink farms led to the euthanising of tens of thousands of mink. Clearly, identifying a particular animal species as being infected, and particularly with the associated worry about future implications for human health, led to a poor animal welfare outcome for many individuals of that species.

3.2.3 Animals found to be infected with SARS-CoV-2 in the wild

It is not only animals that are obviously in close contact with humans which have been found to be SARS-CoV-2 positive. Even animals in the wild have been found to be infected.

Mink were found to be infected in farms, as discussed above, but they have also been found to be virus positive in the wild. An American mink was the first wild animal reported to be positive.[51] Minks in the wild are also vulnerable to other viruses, such as Aleutian mink disease virus (AMDV). Unlike the coronavirus, this is a single-stranded DNA virus, but the symptoms are not that dissimilar. This is typified by interstitial pneumonia, which is usually fatal.[52] The European mink (*Mustela lutreola*) is said to be one of the most endangered mammals in Europe,[53] so a SARS-CoV-2 infection in such populations would not be good news. They are not only threatened by disease, but also by habit loss and competition from the American mink. However, some good news is that vaccinations in captive-bred black-footed ferrets (*Mustela nigripes*) have been shown to be effective,[54] so there is hope on the horizon assuming enough wild animals can be given the vaccine, which will not be an easy job.

The United States Department of Agriculture (USDA) keeps a dashboard of current data on animal SARS-CoV-2 infections.[55] Under the grouping of animals under human care it includes several more species (as of December 2022). These include otters (eight cases) and spotted hyenas (two cases). However, it then lists several species where only one case has been reported. These include binturong (bearcat), coati, cougar, ferret, fishing cat, lynx, mandrill, and squirrel monkey. None of these would be of great surprise, being related to other species known to become infected, including non-human primates, felines, and mustelids. Under wild animals it lists mink, mule deer, and white-tailed deer (Figure 3.4).

It has been established that white-tailed deer (*Odocoileus virginianus*) may be able to be infected with SARS-CoV-2 and also that they have the ability to transmit the virus between individuals, with the authors ending their abstract by saying that their work "identifies white-tailed deer as a wild animal species susceptible to the virus."[56] It was even found that the virus can be transmitted to the foetuses of pregnant animals.[57] And indeed, hundreds of deer in North America have tested positive for the

Figure 3.4 A Chittal deer. Many deer have been reported as being SARS-CoV-2 positive. Painting by Ros Rouse©, although made black and white. The original can be found at https://www.tigerloveart.co.uk.

virus.[58,59,60] Deer of concern were not isolated either, being spread across 24 US states and several provinces in Canada. Furthermore, it is thought that the virus might be mutating in the deer populations. Interestingly, a *Nature* article[61] discussing this says testing was carried out after "teams have cobbled together the funding to survey deer." Therefore, there was no official programme to do this. Also interestingly, working with fawns it was shown that the virus could be shed in nasal mucus and faeces,[62] and therefore it was found to be passed on to other fawns in other pens, indicating how it could spread in a population. It was also stated that this was a rather surprising observation, as other ungulates,[63] such as sheep, cows, and goats, have not been found to get infected. However, looking back at Damas et al.'s data in Table 2.1 does suggest that deer might be more susceptible than its related species. And, in fact, cows have been found to be positive for the virus. Experimentally cattle can become infected,[64,65] and this has been reported in the press.[66]

White-tailed deer (*Odocoileus virginianus*) were also the focus of a study by Hale et al.[67] They found, using reverse transcription real-time PCR, that 35.8% of 360 individual animals tested, that is 129 free-ranging animals, were positive for SARS-CoV-2. The viral variants found in these animals were B.1.2, B.1.582, and B.1.596, and they were concerned that these animals, like the mink, might be reservoirs for the production of new variants. They reported no transmission to humans, but they also pointed out that there is "an urgent need to establish comprehensive "One Health" programmes to monitor the environment, deer and other wildlife hosts globally." Others have also reported SARS-CoV-2 in wild deer. In Québec, Canada, deer were found to be infected with the AY.44 variant of SARS-CoV-2, which is a sublineage of B.1.617.2.[68] Worryingly, it is thought that the virus can evolve in deer at a rate that is three times faster than the mutation rate estimated for the virus in humans,[69] although the reasons for this are unclear. The authors suggest that it may be accounted for by differences in host metabolism, RNA editing enzymes, and many other factors. Indeed, differences in viral evolution rates between host species is widely reported, raising the question of whether this should be studied in all animal populations with significant SARS-CoV-2 infection in order to identify possible sources of novel variants.

In November 2022, the first evidence of deer-to-human transmission of the SARS-CoV-2 virus was reported in *Nature Microbiology*.[70] Clearly, this is an animal species, and human interaction with it, that needs to be carefully monitored during the COVID-19 pandemic and for future similar virus outbreaks.

Cats have been a potential problem, to a degree in those kept as companion animals, and for those larger species kept in captivity, but there have been reports of infected animals in the wild too. For example, a free-ranging Indian leopard (*Panthera pardus fusca*) was found to be infected in India. The animal was infected with the B.1.617.2 variant of the virus.[71] In a similar manner, free roaming dogs (domestic) were found to be infected in Ecuadorian Amazonia.[72] Similarly, stray domestic cats have been found to be SARS-CoV-2 positive in Zaragoza, Spain.[73]

Although there seemed to be little evidence of a problem in the wild, North American deer mice (*Peromyscus maniculatus*) are reported to be able to transmit the virus between individuals,[74] so some monitoring of this species may be prudent. Others have stated that these animals: "have the potential to serve as secondary reservoir hosts in North America."[75]

According to an article in *National Geographic* in January 2023,[76] when discussing wild animals, in the United States minks, mule deer, and white-tailed deer have been reported to be virus positive whilst elsewhere in the world there have been reports of infected black-tailed marmosets, big hairy armadillos, and a leopard. The marmoset, a free-ranging adult female black-tailed marmoset (*Mico melanurus*), was found after a car accident in Cuiabá, Mato Grosso State, Brazil.[77] Four big hairy armadillos (*Chaetophractus villosus*) were found to have been infected in Argentina with the Delta variant of the virus, whilst the leopard (*Panthera pardus fusca*) was found in India, as previously mentioned. The tone of the article suggests that the problem is bigger and more widespread than originally thought, a sentiment they say is shared by Joseph Hoyt at Virginia Tech.

On the other hand, it is important to report what happened in animals which have not been found to be infected. For example, treeshrews, pigs, chickens, turkeys, ducks, geese, and Japanese quail have all been reported as not being able to be infected.[78] However, much of the literature has to be considered together and a balanced view taken, as many papers report contradictory results, especially those dating from the earlier days of the pandemic. For example, it was stated, in amongst discussion of other animal species, that treeshrews were permissive (allowing infection) but dogs were not.[79] For treeshrews this is opposite to the other report cited here, and we know that dogs are able to be infected.

Although there seems to be no reported deaths of marine mammals, there is no lack of concern about this group of animals. For example, work in Italy has highlighted the concerns for marine animals during the COVID-19 pandemic. Wastewater management and extreme weather means that marine mammals are likely to be exposed to the virus. It is known that genetic analysis of the ACE2 protein from many such mammals indicates that they may be susceptible to SARS-CoV-2, and therefore this report highlights that conditions around the likely location of marine mammals means that they are at risk when swimming and feeding in specific areas.[80]

Another example of the downside of the pandemic for marine mammals is shown by the lack of monitoring and retrospective examination of the deaths of manatees in Florida. The Florida Fish and Wildlife Conservation Commission is therefore unable to have robust figures about manatee deaths and the causes, making conservation and management more difficult.[81]

On a more positive note for marine mammals, seal hunting in Canada was virtually stopped during the COVID-19 pandemic. According to the data from Canada's Department for Fisheries and Oceans (DFO), it appears that the overall killing of these animals is less than 8% than normal.[82]

3.2.4 Brief roundup of animals that have been infected

It is clear that many species of animals can become infected by SARS-CoV-2, and some are in close contact with humans, but others are in the wild. In some species animal–animal transmission is seen and therefore there is a potential risk of uncontrolled spread of the virus, especially in the wild, with the concomitant risk of the emergence of new variants and infection of humans. This is not to belittle the suffering and death that may have gone unnoticed in the wild. A summary of some of the data gathered can be found in Table 3.1.

Some researchers have looked at the problem of the virus infections in animals in more detail. Islam et al. looked at the diversity of the viral variants found in animals, both domestic and wild.[83] They list the common mutations seen in the virus from animals, and they seem to be partly species specific. For example, they say:

> the dog was affected mostly by clade O (66.7%), whereas cat and American mink were affected by clade GR (31.6 and 49.7%, respectively).

The authors also recommend vaccination of animals and better genetic surveillance of the virus from animals (see Chapter 4).

Sharun et al.[84] give a useful world map of where animals have become naturally infected by SARS-CoV-2, and which animals were found in each country (Figure 1 in that paper). However, with no criticism of the authors, there are large swathes of the world not represented and therefore one can only assume that this is because of a lack of surveillance, monitoring, and reporting, despite the fact that they sourced their information from the World Organisation for Animal Health (OIE). Rather alarmingly, China is not listed, and there is no doubt that there are naturally infected animals in that region of the world. Delahay and colleagues[85] agree that collection of the data is difficult, and they state:

> Owing to the many data gaps and deficiencies in our understanding of the epidemiology of SARS-CoV-2, it is essential that uncertainty and variability are captured within any risk assessment framework and communicated.

They then give a useful "conceptual framework" (Figure 2 of their paper) for surveillance and identification of particular animal species that need to be the focus of monitoring and reporting to the OIE.

Cui and colleagues also give an excellent overview, published in July 2022, of the animals that have been reported to be SARS-CoV-2

Table 3.1 Examples of animals that were reported to be infected by SARS-CoV-2. Data from a range of sources, including Cui et al. (2022)

Animal	Country	Animal status	Comments
Gorilla	USA Netherlands	In captivity	Mild disease, no deaths reported
Cat	Across at least 27 countries	Companion animal	Usually mild symptoms, can have animal to animal transmission
Cat	Netherlands	Feral	Infection from mink farms
Dog	Across at least 18 countries	Companion animal	Usually mild symptoms
Hamster	Hong Kong	Companion animal	Led to a cull
Lion	India, Spain, South Africa, USA, Sweden, Czech Republic, Estonia, Sri Lanka, Singapore	In captivity	Some deaths
Tiger	USA, Czech Republic	In captivity	
Snow leopard	USA	In captivity	Died of COVID-19 induced complications
Cougar	USA	In captivity	
Puma	USA, Argentina, South Africa	In captivity	
Linx	USA	In captivity	Listed on USDA database
Mink	Netherlands, Denmark, USA, Spain – at least 14 countries in total	Farmed	Deaths, culling, worry of transmission to humans
Mink	USA, Spain	In wild	Worry as endangered
Ferret	USA, Spain, Slovenia	Companion animal	
White-tailed deer	USA, Canada	In wild	Exact numbers not known
Mule deer	USA	In wild	

(Continued)

Table 3.1 (Continued) Examples of animals that were reported to be infected by SARS-CoV-2. Data from a range of sources, including Cui et al. (2022)

Animal	Country	Animal status	Comments
Cattle		Experimental conditions	
Black-tailed marmoset (*Mico melanurus*)	Brazil	In wild	Found after a car accident
Big hairy armadillo (*Chaetophractus villosus*: armadillo peludo)	Argentina	In wild	Four animals found, Delta virus
Binturong (bearcat)	USA	In captivity	Listed on USDA database
Hippopotamus	Belgium	In captivity	
Coatimundi (coati)	USA	In captivity	Listed on USDA database
Fishing cat	USA	In captivity	Listed on USDA database
Hyena	USA	In captivity	Listed on USDA database
Otter	USA	In captivity	Listed on USDA database
Otter (*Lutra lutra*)	Spain	In wild	Eurasian river otter near a water reservoir[i]
Mandril	USA	In captivity	Listed on USDA database
Squirrel monkey	USA	In captivity	Listed on USDA database
Manatee (*Trichechus manatus manatus*)	Brazil	Wild, but managed: Rehabilitation captivity	2 out of 19 animals tested were positive

[i] Padilla-Blanco, M., Aguiló-Gisbert, J., Rubio, V., Lizana, V., Chillida-Martínez, E., Cardells, J., Maiques, E. and Rubio-Guerri, C. (2022) The finding of the severe acute respiratory syndrome coronavirus (SARS-CoV-2) in a Wild Eurasian River Otter (*Lutra lutra*) highlights the need for viral surveillance in wild mustelids. *Frontiers in Veterinary Science, 9*.

positive.[86] They group the animals as farm, captive, pet, and wild, and then they give a world map of where animals were reported as being positive. The map lists 39 countries, although as with other such overviews there is little from Africa and nothing from Australia and New Zealand. They list 18 animal species in their paper, but, interestingly, they point out that such species are across ten families and four orders (Carnivora, Artiodactyla (even-toed ungulates), Primates, and Rodentia). As with other work, there was no evidence of any of the following being found to be SARS-CoV-2 positive: birds, fish, amphibians, reptiles, or lower animals. Published in late 2022, Rao et al.[87] produced a systematic review of SARS-CoV-2 infections in animals, and no doubt there will be further holistic overviews in the future as more is known and the pandemic wanes.

There is clearly more that can be done to understand which animal species have become infected, and which viral variants have been involved. As we write this, it still remains an evolving picture, but this theme will be revisited in Chapter 7.

3.3 What was not known

The concern about knowing which animals have been found to be SARS-CoV-2 positive is that the vast majority of animals have never been tested, and never will. Even species that are highly likely to be susceptible, such as the non-human primates, are not routinely tested. In the wild this would be impossible, and it may even cause issues about infected humans coming into contact with the animal communities. If a gorilla troop had one member infected, it would very likely infect many more. Like humans, such troops would have both very young and older members, some of which may be vulnerable, and therefore it may be wise for humans to stay away, or at least observe strict biosecurity measures, and, rather than vaccinating, only intervene in cases of active respiratory disease. Careful monitoring is suggested and then, as reported by the *Gorilla Doctors*[88]: "gorillas showing signs of respiratory disease—runny nose, coughing, lethargy—are flagged for more intensive observation." There has also been a call for "more responsible tourism to the great apes,"[89] so that human–gorilla interaction is kept to a minimum.

One of the groups thought to be susceptible, from a molecular biology point of view, are the aquatic mammals, such as porpoises, whales, and dolphins. There is some evidence that sea water may be able to carry the virus, but the overall situation is not clear.[90] However, with the discharge of effluent straight into the sea and rivers in many places in the world, it would not be unlikely for aquatic mammals to come into contact with the virus. There seems to be little testing of such animals, although a survey of cetaceans that had been stranded on the Italian coast reported no evidence of infection by SARS-CoV-2 in these animals.[91] Even so, the true scale of the problem, if there is one, remains unknown.

The true scale of the infection of animals we claim to love and cherish is also not known. When Boris Johnston suggested that his dog was tested the idea was derided. However, testing of dogs and cats has never been widely adopted, even though such animals are in contact with a range of humans, not just their immediate owner.

Of course, there is an ethical issue, and one we will revisit in Chapter 4. There is a limit to the amount of testing that can be done. It was a major undertaking ensuring that humans were tested so that the pandemic could be managed. In some parts of the world testing was significantly less efficient (and available) than others, raising concerns about disease spread, as so many humans are so mobile in the 21st century (as well as equalities concerns). Therefore, should limited testing resources be used to test animals when the perceived benefit may be limited? If it could be shown that testing dogs was a good idea, it might have happened. But without that evidence of benefit (either to animals or to humans) the idea of introducing a pet testing programme may have had limited public support, and hence got limited traction (although some pet owners would no doubt have welcomed it). Perhaps that is the correct approach, but on the other hand, if animals could spread the disease and be a possible source of new strains, as seen with the mink farms, then more testing may have been sensible.

Predicting which animals could be infected with SARS-CoV-2 has not been convincing. As discussed in Chapter 2, there are limiting factors about access to the genetic materials needed, but there is also a lack of understanding of exactly why certain animals are more likely to become infected and spread the virus, which ones are likely to be asymptomatic, which will get mild symptoms, and which may die. Why felines and mustelids? What is it about these species that makes them vulnerable and therefore a potential problem? Will the same species be a problem next time (in relation to future, as yet unknown, disease outbreaks) or will other species be more susceptible? Can we predict this?

Of course, one of the major issues, which will be revisited in the last chapter, is what happens in the future. We know that animals, especially bats, can harbour coronaviruses. SARS-CoV-2 was not the first outbreak to hit humans, and it will certainly not be the last. Should more animal testing be a priority to either predict or stop the next epidemic?

3.3.1 Pathways via which evidence emerged

Most of the animals that were found to be SARS-CoV-2 positive were tested because they showed symptoms. Of course, this is a sensible approach; not all animals can be tested on the off chance. Therefore, there were reports of a few companion animals here and there, a few zoo animals, and then some in the wild. Most of these reports were picked up by the popular news outlets, and they were often sensationalised to capture peoples' imagination. This was seen relatively recently in the *Daily Express* in the United Kingdom, for example: "Covid horror as estimated over 350,000 cats infected with virus which 'can be fatal.'"[92] This was an extrapolated number, and it may be accurate, but it is still an estimate. The researchers at Glasgow University looked at 2,309 cats and found that 3.2% of the samples were positive for the virus antibody. With an estimate of the total cat population in the United Kingdom, a number of how many may have been infected was derived. It is a scary number and certainly makes a good headline.

However, national agencies did try to keep track of the spread of SARS-CoV-2 in animal populations. In the USA there was the Centers for Disease Control and Prevention (CDC),[93] and they kept a regularly updated table of the situation in the USA: the SARS dashboard.[94] The data are broken down into groups: companion animals and other animals in human care; premises; wildlife detections. Testing was either by PCR or by antibody detection. At the time of writing (late December 2022), the list included 118 cats, 111 dogs, 55 tigers, 52 lions, 28 gorillas, 13 snow leopards, 8 otters, and 2 spotted hyenas. As discussed previously, there are several other species also listed, including other felines and a mustelid. For animals in the wild, only deer (mule, and white-tailed) and mink are listed. In total 397 animals are listed, which is a long way off the estimated 350,000 cats there might have been infected in the United Kingdom alone. Such numbers highlight the lack of robust evidence and knowledge, and the lack of testing that has taken place in animal populations. However, it does also highlight the spread of the problem as 28 states (out of 50) are thought to have infected animals, but this is almost certainly a major underestimation, although it is right across the USA, from Washington (state) to Florida (there is a schematic map given).

The scientific literature has also attempted to keep up with the current situation. *Scientific American* has a page entitled *Which Animals Catch COVID?* This database has dozens of species and counting.[95] Here there are indeed numerous animals listed, and also their groupings given. Of note on this list, not mentioned here previously, there are eleven cows, seven beavers, two hippopotami, two Caribbean manatee, an anteater, and a black-tailed marmoset.

The finding of manatees (sea cow: *Trichechus manatus manatus*) being positive for SARS-CoV-2 is significant as these mammals live in an aquatic environment, appearing to never leave the water and come on land.[96] Therefore, if they became infected the virus must have come either from the water or from a land-based source, such as intimate contact with humans or perhaps through their food. Animals tested were being looked after by the Brazilian Centre for Research and Conservation of Aquatic Mammals (ICMBio/CMA). Nineteen animals were tested and two were found to be positive. The source of the virus was declared to be unknown, but the Centre altered its biosecurity protocols "to avoid potential human-manatee coronavirus contamination."[97,98] This study shows that other aquatic mammals ought to be monitored more closely. A look back at Table 2.1 will show that many aquatic mammals have been predicted to be highly susceptible to SARS-CoV-2, at least from a molecular biological point of view. However, as with dogs and cats, there are also coronaviruses that are known to infect aquatic mammals, such as in bottlenose dolphins (BdCoV) and beluga whales (BWCoV).[99,100] Vigilance in monitoring this group of animals would be very advisable, albeit not easy or inexpensive to do. On the other hand, recently a new virus which threatens cetaceans was found by targeted surveillance in Hawai'i,[101] so such work is possible (Figure 3.5).

Other databases that collate information about SARS-CoV-2 infections in animals include the Program for Monitoring Emerging Diseases

Figure 3.5 Caribbean manatee (West-Indian manatee or sea cow: *Trichechus manatus*). Photograph from Shutterstock, courtesy of Thierry Eidenweil, but image has been made black and white.

(ProMED),[102] the World Animal Health Information System (WAHIS) of the World Organisation for Animal Health (WOAH, formerly OIE),[103] the Canadian Animal Health Surveillance System (CAHSS),[104] and the Complexity Science Hub (Vienna).[105] The Danish Veterinary and Food Administration also keep a web page about mink infections.[106]

Given the many disparate sources of information listed here, it is not surprising that there has been an attempt to collate all this data together in one place.[107] This has culminated in *A Global Open Access Dataset of Reported SARS-CoV-2 Events in Animals,*[108] abbreviated to SARS-ANI VIS. As of December 2022, there are 741 events reported across 39 countries encompassing 31 animal species. The data allow one look at them by date, so it can be seen if the situation is getting substantially worse. The data are, in fact, relatively straight until it starts to tail off around April 2022. The next graph shows the world spread of the animals reported to be infected. Interestingly there are none in Australia, but most of the rest of the world is involved (it may or may not be relevant that Australia had a zero-COVID policy for much of the pandemic, and it is also a very large country with major population centres dispersed from one another). Felines, canines, and deer are the most obvious species accounting to the majority of animals listed. Hot-tags allows for one to limit the data to fatalities. This is interesting as it lists some that have not obviously popped out from the media. For example, it lists two deaths of dogs in the United Kingdom, four in the USA, one in the Netherlands, one in China, and one in Argentina. For cats, deaths have been reported in the United Kingdom (1), USA (6), Mexico (2), and Brazil (1). For big cats, there are several deaths listed, including four snow leopards and a tiger in the United States, a tiger in Sweden, and a tiger and a lion in India. There is also a squirrel monkey as a listed death in the United States. The next figure gives information about clinical symptoms, and these range from subclinical to death, including sudden death. Nasal discharge is listed for dog, cat, lion, and hippopotamus (this was an animal in Belgium: in fact, a case study report on the infection of two hippopotami in Zoo Antwerp in November 2021 was published[109]). Many animals were listed under respiratory disorders. However, other listed symptoms include myocarditis, gastrointestinal, neurological (including depression), weight loss, and collapse. Clearly there are a range of symptoms seen across animal species, and some species, for example dog, are represented across the spectrum of symptoms.

The next set of data at the SARS-ANI VIS platform is which virus variants have been found in which animal species, and the proportion of animals that have been reported with each. For minks these are all B.1 variants, for example, B.1.1.73 and B.1.1.39 (which accounts for the majority of cases). For cats the viral variants are the GR clade, Delta

AY.3, and B.1.1.7. Half the dogs had B.1.1.7, and the rest were positive for B.1.1.529 and B.1.160. Lions were all B.1.617.2. Variants found in animals as of December 2022 are listed in Table 3.2.

The penultimate section of the SARS-ANI VIS data looks at what control measures were used to try to protect the animals. Interestingly, it is a hot-link figure, so one can look at the outcome and see which

Table 3.2 SARS-CoV-2 variants in animals as listed at SARS-ANI VIS in December 2022

Animal species	SARS-CoV-2 variant reported
American mink	B.1.618
	B.1.173 (close to B1.1)
	B.1.1.72
	B.1.1.73
	B1.1.39
White-tailed deer	B.1.582
	B.1.596
	B.1.617.2
	B.1.311
	B.1.1.529
	B.1.2
	Ontario WTD lineage
Cat	GR clade
	Delta AY.3
	B.1.1.7
Dog	B.1.1.529
	B.1.1.7
	B.1.160
	B.1.1.7
Tiger	B1.177.21
	Delta AY.43
Large hairy armadillo	P.1
Lion	B.1.617.2
Golden hamster	Delta AY.127
Eurasian beaver	B.1.617.2
Leopard	B.1.617.2
Mule deer	B.1.617.2
Ring-tailed coati	P.2

control measures were used. For the total 123 animal deaths, for example, the measures included isolation and quarantine, treatment, culling, euthanasia (listed separately), and movement control. Clearly all these failed for these individual animals, which included a range of felines, dogs, minks, and a squirrel monkey. Under "expected recovery," isolation and treatment seemed to be effective, which was mirrored under "recovery." Under "improved condition" movement control was also added, but also "plans for vaccination," which suggests that the vaccination itself was not used. Such data will be useful moving forward, as they will suggest how effective, or not, certain measures are for controlling the SARS-CoV-2 infection in animal populations.

The last set of data on the SARS-ANI VIS site looks at a summary of all the data, by country. Again, it is hot-linked, so more detail can be gleaned. What is interesting is that some countries, such as Greece, only list one species, and that is mink. Other countries, particularly the USA and those in South America, list a range of species.

The data, therefore, have emerged in a variety of ways. Popular media has reported occasional cases, and mainly when it can be made to sound sensational. Some countries have attempted to collate data and make it public, whilst sites such as SARS-ANI VIS have tried to take a holistic approach. How long into the future such data will be maintained and updated is not known, but one assumes that it will continue in the short to medium term, or perhaps until the next virus becomes more of an issue for human health.

3.4 How risk, uncertainty, and ambiguity were handled

The way risk is understood, tolerated, managed, and responded to is complex. Simply getting more evidence does not always lead to one single view of the "right" response.

To better understand how risks were interpreted and handled, it may be useful to consider some conceptualisations of risk, uncertainty, and ambiguity. Risk is defined as the chance of harm, loss, or hazard. Uncertainty is what frustrates attempts to undertake a probabilistic assessment of risk, because of limited evidence, scientific disagreement, or an acknowledged ignorance.[110,111] Further, Stirling[112,113] has established that ambiguity adds further complexity. Whilst more evidence, data, and information might reduce uncertainty, they will not necessarily reduce ambiguity. Ambiguity is a situation of ambivalence in which different, even divergent, streams of thinking and conceptualisation exist in the consideration and interpretation about a single risk phenomenon. Further, there are two types of ambiguity that may drive different responses. Interpretive ambiguity is where differing groups or individuals take the same evidence and interpret what the evidence means in terms of risk in different ways. Normative ambiguity reflects different concepts of what is regarded as tolerable risk.

It might, therefore, be seen that early in the pandemic there were significant conditions of uncertainty. There was limited evidence, and there was not so much scientific agreement, as scientific evidence came together in a patchwork of limited understanding. It could be argued that in relation to animals, we are still experiencing considerable uncertainty. We were certainly in a situation of ambivalence.

During the earlier phases of the pandemic, whether or not you were locked down (or locked in, as was sometimes the case in China) depended on the way that your government interpreted the ambiguous position in relation to the most appropriate pre-vaccine pandemic response, i.e., they were in a condition of "normative ambiguity."

There has been endless discussion about the origin of the virus, and certainly at times China and other countries have appeared to be following a divergent view. There is still no clear evidence of the precise origin of the virus, although the range of possibilities has contracted over time.

In relation to animals, the concept of "interpretive ambiguity" also comes into play. Looking at the uncertain state of evidence, where an animal (of uncertain species) may have infected, or been infected by, humans at a "wet" market in Wuhan, the way that this ambiguous situation is interpreted could have highly significant implications, the most obvious here being whether or not wet markets should be permitted. Cultural contexts will come into play. Countries, institutions, groups, and individual actors which value such markets[114] are at odds with others who regard them as a biosecurity risk, not to mention a perceived welfare "'horror," for example the view expressed by the People for the Ethical Treatment of Animals (PETA) during the pandemic.[115] Others, including supranational organisations, may situate themselves between these two poles and interpret this situation as being more amenable to improved governance and regulation rather than a ban (such as WHO[116]) or in a clear distinction being made between those wet markets that sell live animals and those that do not.[117]

The public information sources in relation to animals, especially during the earlier phases of the pandemic, were not always helpful in enabling individuals to make risk judgements. Information widely available related to the occasional media report where there was a newsworthy story, such as the odd zoo animal becoming infected, or even dying, which might have been interesting, but had little direct effect on peoples' lives. At the height of the pandemic, zoo and conservation sites were closed anyway, so there was no risk to visitors. There was more risk to the animals if visitors were let back in. Zookeepers were obviously keeping aware of the situation and doing their best to not pass the virus to any animals. The use of PPE was high in such places, but even so some animals were infected.

In human communities, at least in the United Kingdom, dogs were still allowed to mingle, cats were still let out to roam as usual, and there were no restrictions on animal movements. No one seemed to care much, although animal organisations, such as the Royal College of Veterinary Surgeons (RCVS)[118] and the Kennel Club,[119] did give guidance for owners who were concerned and wished to be informed. What information there was focussed on avoiding pets being infected by COVID positive owners and reassuring people that they were unlikely to get COVID from their pet. Academics did research on the topic, data were collated, but the scale of the issue is only now coming to light. Even now (end of 2022) the extrapolated data, such as released for cats in the United Kingom, well outstrips the real data collected, suggesting the reported infections are only the tip of the iceberg.

One of the issues with such data is whether there is a robust story to tell. Most of the time the stories were, as predicted, humancentric: will an infected animal pass the virus back to a human, and will that virus have mutated in the animal? It seemed that unless this was established, there was little interest in the effect on the animals. After all, few were dying compared to humans. In most cases the animals had mild symptoms and would recover. Did it matter if a cat coughed for a few days if the companion human was safe? Local control measures were used where necessary, and this was generally accepted as an appropriate response. Most of the research was on understanding the disease in humans, and the animal story remained, and remains, somewhat unclear due to lack of evidence. There has been limited interest in what was happening to other species, except as a passing concern in response to media reports of danger to wild populations of endangered animals, such as gorillas (a threat that remains, even now the pandemic is officially "over"). There was no groundswell of public pressure for more animal surveillance, at a time when all eyes were focussed on the disease in humans. Media reports focussed on the mutations of the virus as they emerged in humans, and the "animal" dimension focussed on the original source and only occasionally on mutations being found to arise from animal infections (such as mink). Had the expression of the disease in animals been more severe, and/or a clearer link made between infected companion or wild or semi-feral animals passing the disease on to humans, significantly greater systematic efforts would have been needed to produce the necessary evidence. Putting it bluntly, had rats, cats, or birds been found to be a major factor in the mutation of disease and transmission of the virus to humans and other animals, then the degree of surveillance would have needed to be significantly greater. This is an important lesson to learn for any future major disease outbreak or pandemic.

3.5 How can transmission of the virus in animals be better tracked and real-time evidence made publicly and globally available?

With the SARS-CoV-2 pandemic waning (at least in most countries, but as we write – summer 2023 – it is still a major issue in China), it may be time to reflect on what we have learnt and what we can do better. What is obvious from the data above is that it is hard to predict which animals are likely to be involved. Non-human primates will always need a close look in future, keeping an eye on early symptoms, and then treatments or isolation. Other species need better management, a theme we will return to when One Health is discussed later. Certainly, keeping mink in tight cages, where any virus or bacteria, not just SARS-CoV-2, may easily spread seems like a thing to avoid in the future, but commercial pressure may dictate otherwise.

However, as already mentioned, testing and tracking of animals were very limited during the COVID-19 pandemic. At the early stages of any future pandemic (i.e., before it even reaches the stage of being formally designated as such) it would be prudent to consider the role of animals, and not just as an initial source. The significant interest in this pandemic about the source was conflated with geopolitical and philosophical dialogues – was this leaked from a lab in China or was this the result of wet markets selling live wild animals, neither of which reflected especially well on China? Or as the BBC recently put it[120]:

> Those two alternatives now find themselves at the heart of a geopolitical stand-off, a swirling mass of conspiracy theories, and one of the most politicised and toxic scientific debates of our time.

One might argue that better animal surveillance should be put in place for any future epidemic. On the other hand, the costs are huge. Track and trace in the United Kingdom, just for humans, cost £37 billion over two years.[121] To expand this to all animals would realistically never happen. Therefore, targeting will be required, likely involving targeting of some locations where disease outbreaks are taking place as well as targeting specific animal species. In relation to future coronaviruses (including any future SARS-CoV-2 variants of concern) that may mean looking at companion animals, especially dogs and cats, that have been shown to become infected with SARS-CoV-2 and may be expected to become infected with future coronaviruses. They may pass the virus between family members or even between households. Anecdotally, this was thought to be worth exploring in the United Kingdom, but nothing was pragmatically carried out to investigate this. In zoos non-human primates and big cats would be worth targeting for surveillance (which would be likely to happen as there are global networks of zoos, such as the World Association of Zoos and Aquariums (WAZA)[122] and in many countries zoos are heavily regulated), but unless deaths of a particular species were reported in the wild, it would be hard to justify a full-scale global testing programme. In this context, it is interesting to consider the current outbreaks of Avian influenza in mammals.[123,124] In the United Kingdom, the Department for Environment, Food and Rural Affairs (DEFRA) is monitoring the situation, but this will not be the case in all countries in the world where such outbreaks are occurring, especially where countries lack the resources for systematic monitoring. At present, the main mechanism of infection of mammals appears to be predation upon infected birds. However, as Parums has said:

There is now increasing concern that HPAI viruses detected in farmed and wild mammal species in Europe and North America in late 2022 showed molecular adaptation markers for the ability to replicate in mammals.[125]

Parums also notes that there have been concerns relating to infection with the HPAI virus variant A(H5N1) at a mink farm in Galicia, Spain. Finally, Parums also notes:

In 1918, one-third of the world's population, or 500 million people, were infected by the H1N1 virus with avian origin genes, resulting in at least 50 million deaths [1]. The 1918 H1N1 virus also spread between wild birds and poultry before infecting humans.

[note: ref 1[126]]

Whilst the risk to humans is currently being stated as low (by, for example, the WHO, in the United States by the the CDC, and in the United Kingdom by DEFRA), there does appear to be a significant justification for careful monitoring of wild and farmed animals in this regard to prevent serious outbreaks or even a pandemic. Our experience of COVID-19 suggests that in the face of pandemic influenza, we would do well to ensure that there is appropriate animal as well as human surveillance – with a potentially devastatingly high death rate of around 50% it will be critical to understand whether your cat (which may well have predated on birds) can give you the flu.[127] And, or course, if that proves to be the case, and the animal in your home proves to be a real tangible risk to you, what happens? The animal welfare issues here would be suddenly writ large.

What would help in the future would be better and more holistic reporting. People need to know the facts, and the scientific evidence, and what that scientific evidence suggests that they should in practice do in their particular circumstances. It would be essential to stop any panic. If people thought that their pet was a serious danger to them or family members, it is not hard to imagine a mass abandonment of animals, and that would be counterproductive. Whilst for some pet owners the welfare of their pets would be a primary concern, others may put the welfare of themselves and their families first. One study[128] has found that although owners bonded strongly with their pets during the pandemic, this did not alleviate stress and loneliness, and that the relationship between pet ownership and mental health was a complex one, and this complexity would be an important context for the way people may navigate through a situation where their pet was a source of both support as well as stress. In a pandemic where the disease presented a significant risk to life, and where companion animals could be infected, suffer severe disease, and die, the stresses of pet ownership might include concern for the welfare of the pet in terms of infection, concern for the practicalities (and afford-ability) of feeding and caring for the animal in conditions of reduced availability of veterinary care, difficulties in obtaining food in lockdown situations, real or imagined concerns over potential forced culling (and potential subversion of that). And set against this backdrop of increased stress, it would be important to understand how pet owners could be best supported to make sound decisions, for themselves and their animal, as well as for the wider society (see earlier references to decision making under conditions of uncertainty and ambiguity). The alternative to this is not an attractive prospect. As discussed later, animal ownership may have increased during the COVID-19 pandemic. What is not needed is a sig-nificant increase in feral dogs and cats in communities (in terms of either their welfare or their infectivity potential to other animals domestic and wild and to humans). Sound information, which included clear welfare advice, would be critical to properly manage the situation for the benefit of both humans and animals. It is also possible that governments may step in and implement mass culling (as they do currently in cases of disease such as foot and mouth), and in animal loving countries, it is possible this may add a lack of compliance or even a degree of civil unrest to the mix. Equally, if wild animals (such as birds) were found to be presenting a significant risk to human health (such as spreading a disease from which half of those who get it die) the impacts for those populations of wild animals would be profound. In domestic poultry, mass culling has taken place in relation to Avian influenza, but this is about protecting the industry more than it is about protecting people (as the risk to public health is still deemed relatively low). It is conceivable that not only will infected birds be "falling from the sky,"[129] potentially in numbers which will make populations of birds in affected locations extinct,[130] but in a situation where the disease become prevalent in all species, wild birds would be subject to formal (governments adopting formal culling programmes), and informal (citizens taking the law into their own hands) culling measures. If Avian influenza becomes prevalent in wild mammals, then this situation may be extrapolated to (in the UK context) badgers, foxes, hedgehogs etc., and in an international context, the impacts on already pressured wild populations in terms of the disease burden itself, and "culling" in all its forms. This would include the poaching context, where anti-poaching efforts may have less local support, and be less possible in resource terms where the human disease burden was significant. Given the likely crippling financial consequences of a global pandemic the financial resources available for anti-poaching efforts would likely reduce (as was seen in the COVID-19 pandemic, including the impact of the loss of tourist revenue). It was notable that although poaching decreased during the COVID-19 pandemic, primarily due to lockdowns and restrictions on freedom of movement, these markets "bounced back" following those restrictions being removed. However, the consumption of bushmeat increased, due to hardship, and human animal conflict may also have increased.[131] This would also be a fluid situation – at the outset, there may be culling efforts, but if a disease with a 50% human death rate became widely prevalent in human populations, it could be anticipated that there would be a degree of breakdown of civil society, and the consequences of this would have impacts for animals, as well as, obviously, for humans. Having a pandemic preparedness framework in place to cover such issues, including all aspects of the "animal side," for a future pandemic would be critical.

In the United Kingdom, the UKHSA Advisory Board: preparedness for infectious disease threats (updated 2 February 2023)[132] says:

> Zoonotic infections, particularly related to influenza, require collaboration across human and animal sectors, with the sharing of data and research, and through academic partnerships (although the core focus of this report is human health).

Individual country efforts (some better resourced than others) and global institutions are all part of the current system of prevention, detection, and preparedness. Hopefully, global sites such as SARS-ANI VIS will be developed to enable data to be accessible, understandable, and useful for any future epidemics/pandemics. This sort of approach should be able to be adopted for any future viral/bacterial outbreaks so it will be possible to see what impact any future pathogens have on animal species, either in homes, in captivity, or in the wild.

3.6 Non-mammals and COVID-19

There seems to be little doubt that non-vertebrates are not becoming infected with SARS-CoV-2. There have been no reports that any birds, reptiles, fish, or amphibians have been found to be infected, had symptoms, or died as a suspected COVID-19 disease.

Does this mean that all these animals are immune?

An intriguing paper was published in October 2021 by Ives Charlie-Silva and his colleagues from Brazil.[133] They were not looking for a direct infection of animals but what would happen if animals came into contact with fragments of the virus. They created a series of peptides based on the short sections of the spike proteins of SARS-CoV-2, what they referred to as SARS-CoV-2 spike protein peptides (PSDP). They then exposed tadpoles *Physalaemus cuvieri* (a frog found in South America) to various concentrations of three peptides. They found that two of the peptides had effects at both concentrations that they used. Oxidative stress markers were increased in the animals, as were levels of antioxidant enzymes (catalase and superoxide dismutase (SOD)). Acetylcholinesterase activity was increased by all the peptides added, and further *in silico* investigations suggested that these peptides could interact with these enzymes. It is worth pausing over this paper. There is no indication of a COVID-19-like disease here, but breakdown products of the virus do seem to be having biological effects. And this is not in a mammal but in an amphibian, a group of animals being ignored as not being susceptible to SARS-CoV-2. At the end of the abstract to their paper, Ives Charlie-Silva and colleagues say: "These findings indicate that the COVID-19 can constitute environmental impact or biological damage potential." The animals were exposed to 100 and 500 ng/mL of peptide for 24 hours, and this was based on work of others to predict levels of virus to which animals might be exposed. Jamie Shutler and colleagues suggest that the virus can survive in water for up to 25 days, and in highly infected areas the viral load in water can be as much as "infectious doses >100 copies within 100 ml of water."[134] Others have also tried to assess the levels of SARS-CoV-2 in water, with Laura Guerrero-Latorre and colleagues[135] suggesting levels could be as high as 3.19E+06 GC/L (genome copies per litre).[136] when discussing the levels of SARS-CoV-2 in natural water systems, stated:

However, the survival period of coronavirus in water environments strongly depends on temperature, property of water, concentration of suspended solids and organic matter, solution pH, and dose of disinfectant used.

Clearly there is much work that needs to be carried out here before a full understanding of how the COVID-19 virus survives in waste and natural waters, what levels it reaches, and whether breakdown products do have a significant effect on the environment. Perhaps non-vertebrates are not as immune as we think, but, on the other hand, perhaps they have appropriate stress responses that enable them to deal with such environmental contamination. After all, SARS-Cov-2 is not the only virus that would have been in water, whether that is lakes, rivers, or the sea, and animals have had millions of years of evolution to deal with them.

3.7 Conclusions

A look through some of the data in this chapter makes evident some information that is perhaps obvious to predict. Non-human primates, such as gorillas, sometimes live in close proximity to humans, especially in zoos and conservation areas. Because these primates possess a genome so close to that of humans, it is not a great leap to suggest that SARS-CoV-2 will infect these animals, and, indeed, this is what has been seen. But not all the non-human primates seem to be represented above. No chimpanzees were mentioned for example. Why not? It is a good thing, but perhaps a bit of a puzzle. One squirrel monkey was listed, but what of all the other related species?

Looking towards animals that are evolutionarily distant from us, we can see if fish, birds, amphibians, reptiles, and other related vertebrates were affected, and, luckily, none of these species are listed in any of the reports above. Some odd media articles list birds, for example pigeons,[137] but there seems to be no concrete evidence for this. With a huge global industry based around chickens, this is a good thing for food security, although other animal pathogens can have an influence here, such as avian influenza (bird flu).[138] Therefore, there is a range of animals that we do not seem to have to be concerned about when it comes to SARS-CoV-2, which is, of course, a relief and excellent outcome. No predictions suggest that these animals are in need of watching, and no real-life data suggests otherwise (Figure 3.6).

However, there are a range of animals that have been affected. Amongst the companion animals, cats seem to be the species most prominently affected. With estimates that 3.2% of the cat population in the United Kingdom may have been infected, this is a big problem for these animals, although cats that have become infected seem to have mild symptoms.

Figure 3.6 Chickens are often kept in cramped quarters where disease spread is likely. Photograph from Shutterstock, courtesy of Guitar photographer, but has been made black and white.

On the other hand, some felines have died, including lions in zoos. Dogs are also infected, but here the symptoms are less severe than in cats and few deaths have been reported. For example, in July 2020 a seven-year-old German Shepherd, called Buddy, was reported to have died of COVID-19,[139] although the dog may have had lymphoma.[140] Of course, many people have dogs for pragmatic use: seeing-dogs; hearing-dogs; welfare animals. Therefore, such animals may need to be checked and monitored and perhaps vaccinated in future epidemics/pandemics.

Probably the biggest issue has been mink. Being in close quarters with each other and humans, these animals seem to be readily infected – they spread the virus and even seem to be able to transmit the virus, perhaps in a mutated form, back to humans. Other mustelids, e.g., ferrets, were also infected. As discussed below, this has led to millions of indirect deaths of animals. The infection of mink was not predicted, and we think came as a surprise to many, including those working in the mink fur industry.

However, animals in the wild have also been infected. This has included wild mink, which may have been expected once the problem at mink farms had become known, but also in species such as deer (mule and white-tailed). This has been quite widespread across the countries of North America. Many of the animals were found to be positive once killed, as hunting continued. But the scale of the problem is probably not yet fully known, despite the testing programmes that people are trying to put in place.

COVID-19, more correctly referred to as a SARS-CoV-2 infection, is certainly not limited to humans. It has been a major problem, causing millions of human deaths across the globe. Perhaps a few animals becoming infected and a hundred or so animals being reported to have died pales into insignificance when in comparison to the human suffering. That might be the case, but we need to remember that we do not live in isolation, we live in a complex ecosystem. So the following questions may need to be considered to inform future serious disease outbreaks or pandemics:

- Did companion animals help spread the virus? We may never know.
- What was the real risk of an animal species being a "fermentation vessel" for new viruses, which may be released in a mutated form? What was the probability that those new virus variants were more dangerous to humans or to other animal species?
- Why could we not predict which animals could become infected?
- Were we lucky that other vertebrates, such as birds, were immune? What would we have done if this had not been the case? Not just for food (e.g., chickens) but with the risk of disease spread, especially as birds can fly long distances. In this context, avian influenza is highly pertinent to our future considerations (as revisited in Chapter 7).
- Will we ever know the full extent of the virus on animals, especially in the wild? We will not know how many animals died as a result of infection.
- How did different groups make risk judgements under conditions of uncertainty and ambiguity at the different stages of the pandemic, what were their trusted sources of advice, and how could this inform us to provide the right conditions for future informed decision making that supports the health and well-being of both humans and animals?

Animals certainly became infected, some died, and vaccines were used as part of the mitigating control measures tried by some people. We will consider this strategy in the next chapter.

Notes

1 The Times: Boris Johnson's sickly dog Dilyn raised Covid alarm in No 10: https://www.thetimes.co.uk/article/boris-johnsons-sickly-dog-dilyn-raised -covid-alarm-in-no-10-dp3d5dzqq (Accessed 14/12/22).
2 Haake, C., Cook, S., Pusterla, N. and Murphy, B. (2020) Coronavirus infections in companion animals: Virology, epidemiology, clinical and pathologic features. *Viruses*, *12*, 1023.
3 Erles, K., Toomey, C., Brooks, H. W. and Brownlie, J. (2003) Detection of a group 2 coronavirus in dogs with canine infectious respiratory disease. *Virology*, *310*, 216–223.

4   The Hindu: Explained | Ancovax - India's first COVID-19 vaccine for animals: https://www.thehindu.com/sci-tech/science/explained-ancovax-indias-first-covid-19-vaccine-for-animals/article65516811.ece (Accessed 25/05/23).

5   Newman, A., Smith, D., Ghai, R.R., Wallace, R.M., Torchetti, M.K., Loiacono, C., Murrell, L.S., Carpenter, A., Moroff, S., Rooney, J.A. and Behravesh, C.B. (2020) First reported cases of SARS-CoV-2 infection in companion animals—New York, March–April 2020. *Morbidity and Mortality Weekly Report, 69,* 710.

6   Guardian: UK cat cull was considered early in Covid crisis, ex-minister says: https://www.theguardian.com/world/2023/mar/01/uk-cat-cull-was-considered-early-in-covid-crisis-ex-minister-says (Accessed 30/05/23).

7   Gov.org: Covid-19 confirmed in pet dog in the UK: https://www.gov.uk/government/news/covid-19-confirmed-in-pet-dog-in-the-uk (Accessed 14/12/22).

8   Medical News Today: Heart inflammation in cats and dogs: SARS-CoV-2 possible culprit. https://www.medicalnewstoday.com/articles/heart-inflammation-in-cats-and-dogs-sars-cov-2-possible-culprit (Accessed 14/04/23).

9   NBC News: Covid linked to heart inflammation in cats and dogs: https://www.nbcnews.com/health/health-news/covid-linked-heart-inflammation-cats-dogs-rcna4319 (Accessed 14/04/23).

10 Ferasin, L., Fritz, M., Ferasin, H., Becquart, P., Corbet, S., Ar Gouilh, M., Legros, V. and Leroy, E.M. (2021) Infection with SARS-CoV-2 variant B. 1.1. 7 detected in a group of dogs and cats with suspected myocarditis. *Veterinary Record, 189,* e944.

11 George, A.A., Venkataramanan, S.V.A., John, K.J. and Mishra, A.K. (2022) Infective endocarditis and COVID-19 coinfection: An updated review. *Acta Bio Medica: Atenei Parmensis, 93,* e2022030.

12 Kamel, M.S., El-Sayed, A.A., Munds, R.A. and Verma, M.S. (2023) Interactions between humans and dogs during the COVID-19 pandemic: Recent updates and future perspectives. *Animals, 13,* 524.

13 Cats Protection: Can cats catch coronavirus (COVID-19)? https://www.cats.org.uk/help-and-advice/coronavirus/faqs-about-coronavirus-covid-19-and-cats (Accessed 14/12/22).

14 Mirror: Cat dies of Covid caught from UK owner as sufferers urged not to cuddle pets: https://www.mirror.co.uk/news/uk-news/cat-dies-covid-caught-owner-23965637 (Accessed 14/04/23).

15 Jaimes, J.A. and Whittaker, G.R. (2018) Feline coronavirus: Insights into viral pathogenesis based on the spike protein structure and function. *Virology, 517,* 108–121.

16 Carpenter, A., Ghai, R.R., Gary, J., Ritter, J.M., Carvallo, F.R., Diel, D.G., Martins, M., Murphy, J., Schroeder, B., Brightbill, K. and Tewari, D. (2021) Determining the role of natural SARS-CoV-2 infection in the death of domestic pets: 10 cases (2020–2021). *Journal of the American Veterinary Medical Association, 259,* 1032–1039.

17 Mallapaty, S. (2022) How sneezing hamsters sparked a COVID outbreak in Hong Kong. *Nature.* https://doi.org/10.1038/d41586-022-00322-0.

18 Yen, H.-L. et al. (2022) Preprint at social science research network. https://doi.org/10.2139/ssrn.4017393 (Accessed 14/12/22).

19 Blaurock, C., Breithaupt, A., Weber, S., Wylezich, C., Keller, M., Mohl, B.P., Görlich, D., Groschup, M.H., Sadeghi, B., Höper, D. and Mettenleiter, T.C. (2022) Compellingly high SARS-CoV-2 susceptibility of Golden Syrian hamsters suggests multiple zoonotic infections of pet hamsters during the COVID-19 pandemic. *Scientific Reports, 12,* 15069.

20 Guardian: Bristol Zoo gorillas move to new woodland home: https://www.the-guardian.com/world/2021/dec/15/bristol-zoo-gorillas-move-to-new-wood-land-home (Accessed 14/12/22).

21 Guardian: Thirteen gorillas test positive for Covid at Atlanta Zoo: https://www.theguardian.com/us-news/2021/sep/11/gorillas-coronavirus-covid-atlanta-zoo (Accessed 15/12/22).

22 National Geographic: Several gorillas test positive for COVID-19 at California zoo—first in the world: https://www.nationalgeographic.com/animals/article/gorillas-san-diego-zoo-positive-coronavirus (Accessed 19/12/22).

23 Casal, P. and Singer, P. (2021) The threat of great ape extinction from COVID-19. *Journal of Animal Ethics*, *11*, 6–11.

24 BBC: Coronavirus: Four lions test positive for Covid-19 at Barcelona Zoo: https://www.bbc.co.uk/news/world-europe-55229433 (Accessed 19/12/22).

25 Fernández-Bellon, H., Rodon, J., Fernández-Bastit, L., Almagro, V., Padilla-Solé, P., Lorca-Oró, C., Valle, R., Roca, N., Grazioli, S., Trogu, T. and Bensaid, A. (2021) Monitoring natural SARS-CoV-2 infection in lions (*Panthera leo*) at the Barcelona Zoo: Viral dynamics and host responses. *Viruses*, *13*, 1683.

26 The New York Times: Lions infected with Covid spur concern over virus spread in the wild: https://www.nytimes.com/2022/01/20/world/africa/lions-covid-south-african-zoo.html (Accessed 19/12/22).

27 Koeppel, K.N., Mendes, A., Strydom, A., Rotherham, L., Mulumba, M. and Venter, M. (2022) SARS-CoV-2 reverse zoonoses to pumas and lions, South Africa. *Viruses*, *14*, 120.

28 The New York Post: Big cats likely sick with COVID-19 at National Zoo in Washington: https://nypost.com/2021/09/18/lions-tigers-likely-infected-with-covid-19-at-smithsonian-national-zoo/ (Accessed 19/12/22).

29 The Daily Mail: Eight LIONS test positive for Covid-19 at an Indian zoo in the first case of its kind in the country: https://www.dailymail.co.uk/news/article-9540377/Eight-LIONS-test-positive-Covid-19-Indian-zoo.html (Accessed 19/12/22).

30 Karikalan, M., Chander, V., Mahajan, S., Deol, P., Agrawal, R.K., Nandi, S., Rai, S.K., Mathur, A., Pawde, A., Singh, K.P. and Sharma, G.K. (2022) Natural infection of Delta mutant of SARS-CoV-2 in Asiatic lions of India. *Transboundary and Emerging Diseases*, *69*, 3047–3055.

31 Mishra, A., Kumar, N., Bhatia, S., Aasdev, A., Kanniappan, S., Sekhar, A.T., Gopinadhan, A., Silambarasan, R., Sreekumar, C., Dubey, C.K. and Tripathi, M. (2021) SARS-CoV-2 delta variant among Asiatic lions, India. *Emerging Infectious Diseases*, *27*, 2723.

32 Mongabay: Sri Lanka zoo lion contracts COVID-19 as reports of animal infections rise: https://news.mongabay.com/2021/06/sri-lanka-zoo-lion-contracts-covid-19-as-reports-of-animal-infections-rise/ (Accessed 19/12/22).

33 CBS News: Cougar, 3 tigers test positive for COVID-19 at Texas wildlife rescue: https://www.cbsnews.com/dfw/news/in-sync-exotics-cougar-3-tigers-test-positive-covid-19/ (Accessed 19/12/22).

34 McAloose, D., Laverack, M., Wang, L., Killian, M.L., Caserta, L.C., Yuan, F., Mitchell, P.K., Queen, K., Mauldin, M.R., Cronk, B.D. and Bartlett, S.L. (2020) From people to Panthera: Natural SARS-CoV-2 infection in tigers and lions at the Bronx Zoo. *MBio*, *11*, e02220–20.

35 Bartlett, S.L., Diel, D.G., Wang, L., Zec, S., Laverack, M., Martins, M., Caserta, L.C., Killian, M.L., Terio, K., Olmstead, C. and Delaney, M.A. (2021) SARS-CoV-2 infection and longitudinal fecal screening in Malayan tigers (*Panthera tigris jacksoni*), Amur

tigers (*Panthera tigris altaica*), and African lions (*Panthera leo krugeri*) at the Bronx Zoo, New York, USA. *Journal of Zoo and Wildlife Medicine, 51,* 733–744.

36 Forbes: Second Lion in Indian zoo dies of Covid-19 and 10 more still being treated for infection: https://www.forbes.com/sites/siladityaray/2021/06/17/second-lion-in-indian-zoo-dies-of-covid-19-and-10-more-still-being-treated-for-infection/?sh=6780ee012a8b (Accessed 19/12/22).

37 CNN: Snow leopards die of Covid-19 complications at Nebraska zoo: https://edition.cnn.com/2021/11/13/us/coronavirus-snow-leopard-deaths-trnd/index.html (Accessed 19/12/22).

38 National Geographic: Did a mink just give the coronavirus to a human? Here's what we know: https://www.nationalgeographic.com/animals/article/coronavirus-from-mink-to-human-cvd (Accessed 1912/22).

39 Oude Munnink, B.B., Sikkema, R.S., Nieuwenhuijse, D.F., Molenaar, R.J., Munger, E., Molenkamp, R., van der Spek, A., Tolsma, P., Rietveld, A., Brouwer, M., Bouwmeester-Vincken, N., Harders, F., Hakze-van der Honing, R., Wegdam-Blans, M.C.A., Bouwstra, R.J., GeurtsvanKessel, C., van der Eijk, A.A., Velkers, F.C., Smit, L.A.M., Stegeman, A., van der Poel, W.H.M. and Koopmans, M.P.G. (2021) Transmission of SARS-CoV-2 on mink farms between humans and mink and back to humans. *Science, 371,* 172–177.

40 National Geographic: What the mink COVID-19 outbreaks taught us about pandemics: https://www.nationalgeographic.com/animals/article/what-the-mink-coronavirus-pandemic-has-taught-us (Accessed 19/12/22).

41 The Economist: Covid-19 ends Dutch mink farming: https://www.economist.com/europe/2020/09/05/covid-19-ends-dutch-mink-farming (Accessed 14/04/23).

42 Rabalski, L., Kosinski, M., Smura, T., Aaltonen, K., Kant, R., Sironen, T., Szewczyk, B. and Grzybek, M. (2020) Detection and molecular characterisation of SARS-CoV-2 in farmed mink (*Neovision vision*) in Poland. *bioRxiv*.

43 Oreshkova, N., Molenaar, R.J., Vreman, S., Harders, F., Munnink, B.B.O., Hakze-van Der Honing, R.W., Gerhards, N., Tolsma, P., Bouwstra, R., Sikkema, R.S. and Tacken, M.G. (2020) SARS-CoV-2 infection in farmed minks, the Netherlands, April and May 2020. *Eurosurveillance, 25,* 2001005.

44 Lu, L., Sikkema, R.S., Velkers, F.C., Nieuwenhuijse, D.F., Fischer, E.A., Meijer, P.A., Bouwmeester-Vincken, N., Rietveld, A., Wegdam-Blans, M.C., Tolsma, P. and Koppelman, M. (2021) Adaptation, spread and transmission of SARS-CoV-2 in farmed minks and associated humans in the Netherlands. *Nature Communications, 12,* 1–12.

45 Pomorska-Mól, M., Włodarek, J., Gogulski, M. and Rybska, M. (2021) SARS-CoV-2 infection in farmed minks – an overview of current knowledge on occurrence, disease and epidemiology. *Animal, 15,* 100272.

46 De Rooij, M.M., Hakze-Van der Honing, R.W., Hulst, M.M., Harders, F., Engelsma, M., Van De Hoef, W., Meliefste, K., Nieuwenweg, S., Munnink, B.B.O., Van Schothorst, I. and Sikkema, R.S. (2021) Occupational and environmental exposure to SARS-CoV-2 in and around infected mink farms. *Occupational and Environmental Medicine, 78,* 893–899.

47 van Aart, A.E., Velkers, F.C., Fischer, E.A., Broens, E.M., Egberink, H., Zhao, S., Engelsma, M., Hakze-van der Honing, R.W., Harders, F., de Rooij, M.M. and Radstake, C. (2022) SARS-CoV-2 infection in cats and dogs in infected mink farms. *Transboundary and Emerging Diseases, 69,* 3001–3007.

48 Sharun, K., Tiwari, R., Natesan, S. and Dhama, K. (2021) SARS-CoV-2 infection in farmed minks, associated zoonotic concerns, and importance of the One Health approach during the ongoing COVID-19 pandemic. *Veterinary Quarterly, 41,* 50–60.

49   WHO: COVID-19 – Denmark: https://www.who.int/emergencies/disease-out-break-news/item/2020-DON301 (Accessed 31/05/23).

50   Bayarri-Olmos, R., Rosbjerg, A., Johnsen, L.B., Helgstrand, C., Bak-Thomsen, T., Garred, P. and Skjoedt, M.O. (2021) The SARS-CoV-2 Y453F mink variant displays a pronounced increase in ACE-2 affinity but does not challenge antibody neutralization. *Journal of Biological Chemistry, 296.* DOI: https://doi.org/10.1016/j.jbc.2021.100536

51   National Geographic: A mink is the first animal in the wild found with the coronavirus: https://www.nationalgeographic.com/animals/article/wild-mink-tests-positive-coronavirus-utah (Accessed 19/12/22).

52   Zalewski, A., Virtanen, J.M.E., Brzeziński, M., Kołodziej-Sobocińska, M., Jankow, W. and Sironen, T. (2021) Aleutian mink disease: Spatio-temporal variation of prevalence and influence on the feral American mink. *Transboundary and Emerging Diseases, 68*, 2556–2570. https://doi.org/10.1111/tbed.13928.

53   My animals: European mink: Critically endangered: https://myanimals.com/animals/wild-animals-animals/mammals/european-mink-critically-endangered/ (Accessed 21/12/22).

54   The Wildlife Society: Black-footed ferret COVID-19 vaccination seems to be working: https://wildlife.org/black-footed-ferret-covid-19-vaccination-seems-to-be-working/ (Accessed 01/06/23).

55   USDA: SARS dashboard: https://www.aphis.usda.gov/aphis/dashboards/tableau/sars-dashboard (Accessed 21/12/22).

56   Palmer, M.V., Martins, M., Falkenberg, S., Buckley, A., Caserta, L.C., Mitchell, P.K., Cassmann, E.D., Rollins, A., Zylich, N.C., Renshaw, R.W. and Guarino, C. (2021) Susceptibility of white-tailed deer (*Odocoileus virginianus*) to SARS-CoV-2. *Journal of Virology, 95*, e00083–21.

57   Cool, K., Gaudreault, N.N., Morozov, I., Trujillo, J.D., Meekins, D.A., McDowell, C., Carossino, M., Bold, D., Mitzel, D., Kwon, T. and Balaraman, V. (2022) Infection and transmission of ancestral SARS-CoV-2 and its alpha variant in pregnant white-tailed deer. *Emerging Microbes & Infections, 11*, 95–112.

58   Chandler, J.C., Bevins, S.N., Ellis, J.W., Linder, T.J., Tell, R.M., Jenkins-Moore, M., Root, J.J., Lenoch, J.B., Robbe-Austerman, S., DeLiberto, T.J. and Gidlewski, T. (2021) SARS-CoV-2 exposure in wild white-tailed deer (*Odocoileus virginianus*). *Proceedings of the National Academy of Sciences, 118*, e2114828118.

59   Palermo, P.M., Orbegozo, J., Watts, D.M. and Morrill, J.C. (2022) SARS-CoV-2 neutralizing antibodies in white-tailed deer from Texas. *Vector-Borne and Zoonotic Diseases, 22*, 62–64.

60   Kuchipudi, S.V., Surendran-Nair, M., Ruden, R.M., Yon, M., Nissly, R.H., Vandegrift, K.J., Nelli, R.K., Li, L., Jayarao, B.M., Maranas, C.D. and Levine, N. (2022) Multiple spillovers from humans and onward transmission of SARS-CoV-2 in white-tailed deer. *Proceedings of the National Academy of Sciences, 119*, e2121644119.

61   Mallapaty, S. (2022) COVID is spreading in deer. What does that mean for the pandemic? *Nature, 604*, 612–615.

62   Palmer, M.V., Martins, M., Falkenberg, S., Buckley, A., Caserta, L.C., Mitchell, P.K., Cassmann, E.D., Rollins, A., Zylich, N.C., Renshaw, R.W. and Guarino, C. (2021) Susceptibility of white-tailed deer (*Odocoileus virginianus*) to SARS-CoV-2. *Journal of Virology, 95*, e00083–21.

63   Ungulates: Members of the clade Ungulata. These are grouped, such as being odd-toed and even-toed, and are usually mammals with hooves. Sheep, cows, camels and deer all fall into the even-toed grouping.

64   Ulrich, L., Wernike, K., Hoffmann, D., Mettenleiter, T.C. and Beer, M. (2020) Experimental infection of cattle with SARS-CoV-2. *Emerging Infectious Diseases, 26*, 2979–2981. https://doi.org/10.3201/eid2612.203799.

65 Bosco-Lauth, A.M., Walker, A., Guilbert, L., Porter, S., Hartwig, A., McVicker, E., Bielefeldt-Ohmann, H. and Bowen, R.A. (2021) Susceptibility of livestock to SARS-CoV-2 infection. *Emerging Microbes & Infections*, 10, 2199–2201. https://doi.org/10.1080/22221751.2021.2003724.

66 Mail online: Cattle, sheep and even cats CAN catch coronavirus: Bombshell report reveals disease can now infect at least ten other animals as well as bats: https://www.dailymail.co.uk/sciencetech/article-8232097/How-cattle-sheep-cats-catch-coronavirus.html (Accessed 21/12/22).

67 Hale, V.L., Dennis, P.M., McBride, D.S., Nolting, J.M., Madden, C., Huey, D., Ehrlich, M., Grieser, J., Winston, J., Lombardi, D. and Gibson, S. (2022) SARS-CoV-2 infection in free-ranging white-tailed deer. *Nature*, 602, 481–486.

68 Kotwa, J.D., Massé, A., Gagnier, M., Aftanas, P., Blais-Savoie, J., Bowman, J., Buchanan, T., Chee, H.Y., Dibernardo, A., Kruczkiewicz, P. and Nirmalarajah, K. (2022) First detection of SARS-CoV-2 infection in Canadian wildlife identified in free-ranging white-tailed deer (*Odocoileus virginianus*) from southern Québec, Canada. *BioRxiv*, 2022–01.

69 McBride, D., Garushyants, S., Franks, J., Magee, A., Overend, S., Huey, D., Williams, A., Faith, S., Kandeil, A., Trifkovic, S. and Miller, L. (2023) Accelerated evolution of SARS-CoV-2 in free-ranging white-tailed deer. ResearchSquare. Available at https://assets.researchsquare.com/files/rs-2574993/v1/232b1ab2-4889-4b8d-b28f-d12108d1e59a.pdf?c=1678218999 (Accessed 13/03/23).

70 Pickering, B., Lung, O., Maguire, F., Kruczkiewicz, P., Kotwa, J.D., Buchanan, T., Gagnier, M., Guthrie, J.L., Jardine, C.M., Marchand-Austin, A. and Massé, A. (2022) Divergent SARS-CoV-2 variant emerges in white-tailed deer with deer-to-human transmission. *Nature Microbiology*, 7, 1–14.

71 Mahajan, S., Karikalan, M., Chander, V., Pawde, A.M., Saikumar, G., Semmaran, M., Lakshmi, P.S., Sharma, M., Nandi, S., Singh, K.P. and Gupta, V.K. (2022) Detection of SARS-CoV-2 in a free ranging leopard (*Panthera pardus fusca*) in India. *European Journal of Wildlife Research*, 68, 59.

72 Zambrano-Mila, M.S., Freire-Paspuel, B., Orlando, S.A. and Garcia-Bereguiain, M.A. (2022) SARS-CoV-2 infection in free roaming dogs from the Amazonian jungle. *One Health*, 14, 100387.

73 Villanueva-Saz, S., Giner, J., Tobajas, A.P., Pérez, M.D., González-Ramírez, A.M., Macías-León, J., González, A., Verde, M., Yzuel, A., Hurtado-Guerrero, R. and Pardo, J. (2022) Serological evidence of SARS-CoV-2 and co-infections in stray cats in Spain. *Transboundary and Emerging Diseases*, 69, 1056–1064.

74 Griffin, B.D., Chan, M., Tailor, N., Mendoza, E.J., Leung, A., Warner, B.M., Duggan, A.T., Moffat, E., He, S., Garnett, L. and Tran, K.N. (2021) SARS-CoV-2 infection and transmission in the North American deer mouse. *Nature Communications*, 12, 1–10.

75 Fagre, A., Lewis, J., Eckley, M., Zhan, S., Rocha, S.M., Sexton, N.R., Burke, B., Geiss, B., Peersen, O., Bass, T. and Kading, R. (2021) SARS-CoV-2 infection, neuropathogenesis and transmission among deer mice: Implications for spillback to New World rodents. *PLoS Pathogens*, 17, e1009585.

76 National Geographic: COVID-19 is more widespread in animals than we thought: https://www.nationalgeographic.co.uk/animals/2023/01/covid-19-is-more-widespread-in-animals-than-we-thought (Accessed 14/03/23).

77 Pereira, A.H., Vasconcelos, A.L., Silva, V.L., Nogueira, B.S., Silva, A.C., Pacheco, R.C., Souza, M.A., Colodel, E.M., Ubiali, D.G., Biondo, A.W. and Nakazato, L. (2022) Natural SARS-CoV-2 infection in a free-ranging black-tailed marmoset (*Mico melanurus*) from an urban area in mid-west Brazil. *Journal of Comparative Pathology*, 194, 22–27.

78 Mahdy, M.A., Younis, W. and Ewaida, Z. (2020) An overview of SARS-CoV-2 and animal infection. *Frontiers in Veterinary Science*, 7, 596391.

79 Abdel-Moneim, A.S. and Abdelwhab, E.M. (2020) Evidence for SARS-CoV-2 infection of animal hosts. *Pathogens*, 9, 529.

80 Audino, T., Grattarola, C., Centelleghe, C., Peletto, S., Giorda, F., Florio, C.L., Caramelli, M., Bozzetta, E., Mazzariol, S., Di Guardo, G. and Lauriano, G. (2021) SARS-CoV-2, a threat to marine mammals? A study from Italian seawaters. *Animals*, 11, 1663.

81 Florida Today: Coronavirus clouds causes of manatee deaths in 2020: https://eu.floridatoday.com/story/news/local/environment/lagoon/2020/12/04/corona-virus-clouds-causes-manatee-deaths/3808350001/ (Accessed 25/05/23).

82 Fur for Animals: Canada's seal hunt has been virtually stopped by Covid-19: https://respectforanimals.org/canadas-seal-hunt-has-been-virtually-stopped-by-covid-19/ (Accessed 25/05/23).

83 Islam, A., Ferdous, J., Sayeed, M.A., Islam, S., Kaisar Rahman, M., Abedin, J., Saha, O., Hassan, M.M. and Shirin, T. (2021) Spatial epidemiology and genetic diversity of SARS-CoV-2 and related coronaviruses in domestic and wild animals. *Plos One*, 16, e0260635.

84 Sharun, K., Dhama, K., Pawde, A.M., Gortázar, C., Tiwari, R., Bonilla-Aldana, D.K., Rodriguez-Morales, A.J., de la Fuente, J., Michalak, I. and Attia, Y.A. (2021) SARS-CoV-2 in animals: Potential for unknown reservoir hosts and public health implications. *Veterinary Quarterly*, 41, 181–201.

85 Delahay, R.J., de la Fuente, J., Smith, G.C., Sharun, K., Snary, E.L., Flores Giron, L., Nziza, J., Fooks, A.R., Brookes, S.M., Lean, F.Z. and Breed, A.C. (2021) Assessing the risks of SARS-CoV-2 in wildlife. *One Health Outlook*, 3, 1–14.

86 Cui, S., Liu, Y., Zhao, J., Peng, X., Lu, G., Shi, W., Pan, Y., Zhang, D., Yang, P. and Wang, Q. (2022) An updated review on SARS-CoV-2 infection in animals. *Viruses*, 14, 1527.

87 Rao, S.S., Parthasarathy, K., Sounderrajan, V., Neelagandan, K., Anbazhagan, P. and Chandramouli, V. (2023) Susceptibility of SARS Coronavirus-2 infection in domestic and wild animals: A systematic review. *3 Biotech*, 13, 5.

88 Gorilla doctors: Gorilla health threats: https://www.gorilladoctors.org/saving-lives/gorilla-health-threats/infectious-disease/ (Accessed 31/05/23).

89 Kalema-Zikusoka, G., Rubanga, S., Ngabirano, A. and Zikusoka, L. (2021) Mitigating impacts of the COVID-19 pandemic on gorilla conservation: Lessons from Bwindi Impenetrable Forest, Uganda. *Frontiers in Public Health*, 9, p.655175.

90 Guo, W., Cao, Y., Kong, X., Kong, S. and Xu, T. (2021) Potential threat of SARS-CoV-2 in coastal waters. *Ecotoxicology and Environmental Safety*, 220, 112409.

91 Audino, T., Berrone, E., Grattarola, C., Giorda, F., Mattioda, V., Martelli, W., Pintore, A., Terracciano, G., Cocumelli, C., Lucifora, G. and Nocera, F.D. (2022) Potential SARS-CoV-2 susceptibility of cetaceans stranded along the Italian coastline. *Pathogens*, 11, 1096.

92 Express: Covid horror as estimated over 350,000 cats infected with virus which 'can be fatal': https://www.express.co.uk/news/uk/1699730/Covid-19-cats-University-of-Glasgow-veterinarians-virologists-Grace-Tyson-ont (Accessed 19/12/22).

93 Centers for Disease Control and Prevention (CDC): Animals and COVID-19: https://www.cdc.gov/coronavirus/2019-ncov/daily-life-coping/animals.html (Accessed 21/12/22).

94 USDA: SARS dashboard: https://www.aphis.usda.gov/aphis/dashboards/tableau/sars-dashboard (Accessed 21/12/22).

95 Scientific American: Which animals catch COVID? This database has dozens of species and counting: https://www.scientificamerican.com/article/which-animals-catch-covid-this-database-has-dozens-of-species-and-counting/ (Accessed 21/12/22).

96 Captain Mike's: Do Manatees ever go on land?: https://swimmingwiththemanatees.com/do-manatees-ever-go-on-land/ (Accessed 14/04/23).

97 Melo, F.L., Bezerra, B., Luna, F.O., Barragan, N.A.N., Arcoverde, R.M.L., Umeed, R., Lucchini, K. and Attademo, F.L.N. (2022) Coronavirus (SARS-CoV-2) in Antillean Manatees (*Trichechus manatus manatus*). ResearchSquare. https://doi.org/10.21203/rs.3.rs-1065379/v1.

98 Attademo, L.N. (2022) Investigation of Coronavirus (SARS-CoV-2) in Antillean manatees (*Trichechus Manatus Manatus*) in Northeast Brazil. *Science World Journal of Cancer Sciences and Therapy*, 1, 1–9.

99 Woo, P.C., Lau, S.K., Lam, C.S., Tsang, A.K., Hui, S.W., Fan, R.Y., Martelli, P. and Yuen, K.Y. (2014) Discovery of a novel bottlenose dolphin coronavirus reveals a distinct species of marine mammal coronavirus in Gammacoronavirus. *Journal of Virology*, 88, 1318–1331.

100 Wang, L., Maddox, C., Terio, K., Lanka, S., Fredrickson, R., Novick, B., Parry, C., McClain, A. and Ross, K. (2020) Detection and characterization of new coronavirus in bottlenose dolphin, United States, 2019. *Emerging Infectious Diseases*, 26, 1610.

101 Clifton, C.W., Silva-Krott, I., Marsik, M. and West, K.L. (2023) Targeted surveillance detected novel beaked whale circovirus (BWCV) in ten new host cetacean species across the Pacific basin. *Frontiers in Marine Science*, 2386.

102 ProMed: International Society for Infectious Diseases: https://promedmail.org/ (Accessed 21/12/22).

103 The World Animal Health Information System (WAHIS) of the World Organisation for Animal Health (WOAH, formerly OIE): https://wahis.woah.org/ (Accessed 21/12/22).

104 Canadian Animal Health Surveillance System (CAHSS): https://cahss.ca/cahss-tools/sars-cov-2-dashboard (Accessed 21/12/22).

105 The Complexity Science Hub (Vienna): SARS-ANI VIS: https://vis.csh.ac.at/sars-ani/#infections (Accessed 01/06/23).

106 Danish Veterinary and Food Administration: https://www.foedevarestyrelsen.dk/Dyr/Dyr-og-Covid-19/Mink-og-COVID-19 (Accessed 21/12/22: a translation is available).

107 Nerpel, A., Yang, L., Sorger, J., Kaesbohrer, A., Walzer, C. and Desvars-Larrive, A. (2022) SARS-ANI: A global open access dataset of reported SARS-CoV-2 events in animals. *bioRxiv*.

108 SARS-ANI VIS: A global open access dataset of reported SARS-CoV-2 events in animals: https://vis.csh.ac.at/sars-ani/ (Accessed 21/12/22).

109 Vercammen, F., Cay, B., Gryseels, S., Balmelle, N., Joffrin, L., Van Hoorde, K., Verhaegen, B., Mathijs, E., Van Vredendaal, R., Dharmadhikari, T. and Chiers, K. (2023) SARS-CoV-2 infection in captive hippos (*Hippopotamus amphibius*), Belgium. *Animals*, 13, 316.

110 Aven, T. and Renn, O. (2009) On risk defined as an event where the outcome is uncertain. *Journal of Risk Research*, 12, 1–11.

111 Filar, J.A. and Haurie, A. (2010) *Uncertainty and Environmental Decision Making*. New York: Springer.

112 Stirling, A. (2003) Risk, uncertainty and precaution: Some instrumental implications from the social sciences. *Negotiating Environmental Change: New Perspectives from Social Science*, 33–76.

113 Stirling, A. (2009) Risk, uncertainty and power. *Seminar Magazine*, 597, 33–39.

114 FOOD52: Wet markets are essential to Thai cooking. So why are they disappearing? https://food52.com/blog/25572-why-wet-markets-are-essential (Accessed 01/06/23).

115 PETA: Filthy 'Wet markets' are still selling scared animals and rotting flesh despite mounting COVID-19 death toll: https://investigations.peta.org/indonesia-thailand-wet-markets/ (Accessed 01/06/23).

116 BBC: Coronavirus: WHO developing guidance on wet markets: https://www.bbc.co.uk/news/science-environment-52369878 (Accessed 01/06/23).

117 CNN: China's wet markets are not what some people think they are: https://edition.cnn.com/2020/04/14/asia/china-wet-market-coronavirus-intl-hnk/index.html (Accessed 01/06/23).

118 RCVS: Key coronavirus guidance updated in light of latest restrictions: https://www.rcvs.org.uk/news-and-views/news/key-coronavirus-guidance-updated-in-light-of-latest/ (Accessed 01/06/23).

119 Kennel Club: Coronavirus (Covid-19): https://www.thekennelclub.org.uk/coronavirus/ (Accessed 01/06/23).

120 BBC: Covid: Top Chinese scientist says don't rule out lab leak: https://www.bbc.co.uk/news/world-asia-65708746 (Accessed 01/06/23).

121 UK Parliament: "Unimaginable" cost of Test & Trace failed to deliver central promise of averting another lockdown: https://committees.parliament.uk/committee/127/public-accounts-committee/news/150988/unimaginable-cost-of-test-trace-failed-to-deliver-central-promise-of-averting-another-lockdown/ (Accessed 21/12/22).

122 WAZA: World Association of Zoos and Aquariums: https://www.waza.org/ (Accessed 01/06/23).

123 UK Government: Coverage on avian influenza in non-avian wild mammals: https://deframedia.blog.gov.uk/2023/02/02/coverage-on-avian-influenza-in-non-avian-wild-mammals/ (Accessed 01/06/23).

124 UK Government: Bird flu (avian influenza): Findings in non-avian wildlife: https://www.gov.uk/government/publications/bird-flu-avian-influenza-findings-in-non-avian-wildlife (Accessed 01/06/23).

125 Parums, D.V. (2023) Global surveillance of highly pathogenic avian influenza viruses in poultry, wild birds, and mammals to prevent a human influenza pandemic. *Medical Science Monitor: International Medical Journal of Experimental and Clinical Research, 29*, e939968–1.

126 Kilbourne, E.D. (2006) Influenza pandemics of the 20th century. *Emerging Infectious Diseases, 12*, 9–14.

127 WHO: Human infection with avian influenza A(H5) viruses: https://cdn.who.int/media/docs/default-source/wpro---documents/emergency/surveillance/avian-influenza/ai_20230512.pdf?sfvrsn=5f006f99_114#:~:text=As%20of%206%20April%202023,(CFR)%20of%2056%25 (Accessed 01/06/23).

128 ScienceNews: Pets and people bonded during the pandemic. But owners were still stressed and lonely: https://www.sciencenews.org/article/pets-owners-bond-pandemic-stress (Accessed 15/08/23).

129 Guardian: 'Falling from the sky in distress': The deadly bird flu outbreak sweeping the world: https://www.theguardian.com/science/audio/2022/jul/19/falling-from-the-sky-in-distress-the-deadly-bird-flu-outbreak-sweeping-the-world (Accessed 01/06/23).

130 Guardian: I helped pick up 6,000 dead birds last summer. This is what I learned about the horrors of bird flu: https://www.theguardian.com/environment/2023/jan/04/farne-islands-bird-flu-i-was-one-in-hazmat-suit-we-picked-up-6000-dead-birds-aoe (Accessed 01/06/23).

131 K4D: Impact of COVID-19 on poaching and illegal wildlife trafficking trends in Southern Africa: https://opendocs.ids.ac.uk/opendocs/bitstream/handle/20.500 .12413/17179/1094_Impact_of_COVID-19_on_poaching_and_illegal_wildlife _trafficking.pdf?sequence=1&isAllowed=y (Accessed 01/06/23).

132 UK Government: UKHSA Advisory Board: Preparedness for infectious disease threats: https://www.gov.uk/government/publications/ukhsa-board-meeting -papers-january-2023/ukhsa-advisory-board-preparedness-for-infectious-disease -threats (Accessed 01/06/23).

133 Charlie-Silva, I., Araújo, A.P., Guimarães, A.T., Veras, F.P., Braz, H.L., de Pontes, L.G., Jorge, R.J., Belo, M.A., Fernandes, B.H., Nóbrega, R.H. and Galdino, G. (2021) Toxicological insights of Spike fragments SARS-CoV-2 by exposure environment: A threat to aquatic health? *Journal of Hazardous Materials*, *419*, 126463.

134 Shutler, J., Zaraska, K., Holding, T., Machnik, M., Uppuluri, K., Ashton, I., Migdał, Ł. and Dahiya, R. (2020) Risk of SARS-CoV-2 infection from contaminated water systems. *MedRxiv*, 2020–06.

135 Guerrero-Latorre, L., Ballesteros, I., Villacrés-Granda, I., Granda, M.G., Freire-Paspuel, B. and Ríos-Touma, B. (2020) SARS-CoV-2 in river water: Implications in low sanitation countries. *Science of the Total Environment*, *743*, 140832.

136 Tran, H.N., Le, G.T., Nguyen, D.T., Juang, R.S., Rinklebe, J., Bhatnagar, A., Lima, E.C., Iqbal, H.M., Sarmah, A.K. and Chao, H.P. (2021) SARS-CoV-2 coronavirus in water and wastewater: A critical review about presence and concern. Environmental Research, 193, 110265.

137 Mail online: Cattle, sheep and even cats CAN catch coronavirus: Bombshell report reveals disease can now infect at least ten other animals as well as bats: https://www.dailymail.co.uk/sciencetech/article-8232097/How-cattle-sheep -cats-catch-coronavirus.html (Accessed 21/12/22).

138 Gov.UK: Bird flu (avian influenza): How to spot and report it in poultry or other captive birds: https://www.gov.uk/guidance/avian-influenza-bird-flu (Accessed 21/12/22).

139 CNET: Pet owners shouldn't panic about the dog that died after COVID-19 infection: https://www.cnet.com/science/pet-owners-shouldnt-panic-about-the -dog-that-died-after-covid-19-infection/ (Accessed 01/06/23).

140 CNN: First dog to test positive for Covid-19 in the US, Buddy the German shepherd, has died: https://edition.cnn.com/2020/07/30/health/first-dog-dies -coronavirus-wellness-trnd/index.html (Accessed 07/06/23).

4

Animal vaccines

4.1 Introduction

When there is a significant impact of a disease in humans, one of the strategies that can be employed is the use of vaccines. Many people are routinely vaccinated against a range of diseases, especially as a child, and often this is carried out at schools[1] (Figure 4.1).

The idea of inoculating against a future disease was developed by Edward Jenner (1749–1823). He was born in Berkeley, Gloucestershire, UK, and he went on to train as a medic before returning to set up his own doctor's practice. He was an interesting person who had eclectic interests. He wrote poetry and was interested in cuckoos.[2] But it was for his work on vaccinations that he was most famous.

Sarah Nelmes, a dairymaid, visited Jenner in May 1796 and he diagnosed her as having cowpox. This obviously made him think about the disease and how it may be related to smallpox. On 14 May he scratched the arm of James Phipps with cowpox. James was an eight-year-old boy, and, in fact, he was the son of his gardener. In July he gave James smallpox to see what would happen. It was obviously a risky thing to do as smallpox is often fatal (and, or course, would be viewed as ethically unacceptable now). James survived, and Jenner realised that giving one pox could protect against another and the idea of inoculating against disease was born.

Today vaccinations are used for a wide range of diseases, so when it was realised that COVID-19 was going to be such a major problem around the world there was a scramble by the relevant laboratories to develop a vaccine that could be used at the scale needed – ideally all humans would need to be vaccinated. The first vaccines developed were

DOI: 10.1201/9781003427254-4

Figure 4.1 Edward Jenner, who pioneered the development of vaccines, including the smallpox vaccine. Illustration by German Vizulis, on Shutterstock. Image has been made black and white.

by Moderna, the University of Oxford in collaboration with AstraZeneca, and CanSino Biologics. A comprehensive timeline of the vaccine developments was published in June 2021.[3]

4.1.1 Types of COVID-19 vaccines

The vaccine for SARS-CoV-2 was not the only anti-SARS vaccine developed. There were vaccines researched for the use against MERS,[4,5] for example, so there was some precedent here. Even so, not all the vaccines available for use in humans were based on the same technology. In fact, there were several platforms used for producing human vaccines,[6,7] as outlined in Table 4.1

One of the oldest ways to make a vaccine is to use a part of one of the proteins that is contained within the virus. A small section of a protein

Table 4.1 Some platforms and vectors that can be used for development of future vaccines

Platform		Comments
Underpinning technologies	Live attenuated virus	Some safety issues
	Whole inactivated virus	Some safety issues and needs adjuvant
	DNA-based	Low levels of immunogenicity; hard to roll out
	mRNA-based	Needs cold storage for rollout. Adaptable for future viruses
Vectors	Recombinant	Can be produced by using cell lines
	Protein based	May need adjuvant and harder to adapt to new viruses

Information drawn from Excler et al. (2021)[8].

that can been "seen" on the outside of the virus is created and then injected into the patient. Usually there is an accelerator agent added, called an adjuvant. This triggers the immune system of the person to react to the viral protein segment present and the body's systems are left with a long-term defence against the presence of that particular protein. If the virus tries to invade, the protein on its surface is rapidly recognised and a swift immune response stops the virus taking hold and spreading. This type of vaccine was first used in the hepatitis B vaccine, and it is also used in the whooping cough (hundred-day cough) vaccines. This was also the basis for the Novavax COVID-19 vaccine.

Viral vectors are also used to create vaccines. Here a second virus (a vector) is used to carry a gene, which will encode for a protein that is in the virus for which the vaccine is needed. The vector enables the protein of interest to be expressed and therefore the immune system will "see" it and react to it. This can then give a long-term immunity against the pathogen of concern. Again, if the pathogenic virus tries to invade, the body has seen one of its proteins before and is ready to mount a swift and efficient defence. This technology has been used for vaccines against Zika, influenza, Ebola, and human immunodeficiency virus (HIV). It is also the basis of the Oxford/AstraZeneca and Johnson & Johnson's Janssen COVID-19 vaccine, for example.

A more recent approach is to use what is known as mRNA technology. mRNA, more fully known as messenger ribonucleic acid, is the molecule that cells make as a copy of the DNA encoding for a protein which needs to be made. The mRNA replicates the information needed to make the protein, and then in eukaryotes (such as animals and plants) the mRNA leaves the nucleus of the cell. Once in the cell's cytoplasm,

Table 4.2 Vaccines with EUL as reported by WHO in January 2022

Vaccine/manufacturer	Date given Emergency Use Listing
Pfizer/BioNTech Comirnaty	31/12/2020
SII/COVISHIELD	16/02/21
AstraZeneca/AZD1222	16/02/21
Janssen/Ad26.COV 2.S	12/03/21
Moderna COVID-19 (mRNA 1273)	30/04/21
Sinopharm COVID-19	07/05/21
Sinovac-CoronaVac	01/06/21
Bharat Biotech BBV152 COVAXIN	03/11/21
Covovax (NVX-CoV2373)	17/12/21
Nuvaxovid (NVX-CoV2373)	20/12/21

large protein structures called ribosomes attach to the mRNA and use the information encoded there to attach the right amino acids in the right order to create a polypeptide (or otherwise referred to as the protein – although this is somewhat simplistic as polypeptides can be heavily modified before working proteins are produced). Therefore, if a mRNA can be produced that encodes for part of a protein from a pathogenic virus, that mRNA can be injected into a person and the cells' systems hijacked to make the protein. Once in the body, that protein will be recognised as foreign, an immune response will be mounted, and immunity established. This technology has been considered for a range of diseases, including influenza, Zika, and rabies, but is the basis of the Pfizer-BioNTech or Moderna COVID-19 vaccines.

WHO lists the vaccines that were given Emergency Use Listing (EUL) for use on humans as of 12 January 2022,[9] and these are listed in Table 4.2.

4.1.2 COVID-19 vaccine rollout and alternate therapies

Once the vaccines were ready there was then a major programme to roll out the delivery to the public. With a population estimated to be 67 million in mid-2021 in the United Kingdom, this was a major undertaking. In the United States, with a population of approximately 330 million,[10] the problem is even bigger. Therefore, governments put in massive resources to make this happen. In many countries, populations were surprisingly rapidly inoculated, but this was not universal across the globe. In August 2021 the World Economic Forum was bemoaning the amount of work still to be done to get people vaccinated, giving statistics for many countries where far less than 10% of the population was only partially vaccinated, and many countries where there was no full vaccination of people shown. The site also correlated the countries' wealth with the percentage vaccination. There is no surprise that the richest countries could get their populations vaccinated, but the poorer countries could not. Some politicians, such as Gordon Brown in the United Kingdom, were very vocal about this disparity,[11] but the situation has sadly continued. This will be further discussed later when we revisit ethics (Figure 4.2).

On 2 December 2020, the Pfizer-BioNTech Covid-19 vaccine was approved for use in the United Kingdom. This was the first such vaccine to be authorised anywhere in the world. It was not long before another was approved. On 30 December the Oxford-AstraZeneca vaccine started to be used and this was both cheaper and easier to distribute. In December 2021 the UK government ordered more than 650m doses of vaccines.[12] These were from eight different sources, and not all were approved for use at the time – it was hoped approval would soon follow.

Vulnerable groups were identified and people were invited to be vaccinated. A similar situation was replicated in many countries, but not all populations and communities were so lucky. Several areas of the world, notably Africa, were left behind as the vaccine roll-out proceeded at pace. As of February 2023, some people in the United Kingdom, including two of the authors of this book, have had four vaccinations as it is feared that

Figure 4.2 Ampoules of SARS-CoV-2 vaccines, ready for use. Photograph from Shutterstock, courtesy of M-Foto. Image has been made black and white.

immunity may not be long lived, especially as new variants of SARS-COV-2, such as Omicron, spread through the populations. There have been numerous variants reported, with some of them being listed in Table 1.2. Data from *Our World in Data* will even break the information down to reveal which countries are affected most by which variants. As of Jan 2023, it shows that the United Kingdom is mainly affected by Omicron variants, whereas in Australia over a third (34.89%) of infections are the result of what is listed as a recombinant variant.[13] As stated by Cosar and colleagues, in the SARS-CoV-2 genome:

> Thousands of mutations have accumulated and continue to since the emergence of the virus.[14]

Not all these mutations will be of any consequence. Within DNA/RNA sequences is a phenomenon referred to as redundancy. This means that there can be a change in the nucleotide sequence but that does not manifest itself in a change in the amino acid sequence in the protein finally produced using that nucleotide sequence. This is because nucleotide-based information is in the form of triplets, called codons, and more than one of these can encode for the use of the same amino acid (as discussed in Section 2.2). For example, the amino acid serine is encoded for six different codons, so if a mutation simply changes the codon to look like another for serine, nothing pragmatically happens. There is further redundancy within the use of amino acids themselves. Many have similar chemical properties, so changing one for the other may have little effect on the structure or function of the protein. For example, changing alanine (which has a methyl side-chain) to a glycine (a proton side change) may make no difference to the protein and spatially and chemically both are very similar. Therefore, numerous mutations in SARS-CoV-2 are of little or no consequence and can be ignored. Headlines which talk about thousands of variants can be misleading. On the other hand, there have been significant variants that have arisen, such as the Omicron variants, which now account for the majority of COVID-19 cases in the United Kingdom and the United States, and new variants could potentially be a problem if the vaccine is no longer effective against them. And as previously discussed, we do not have much robust information about how variants affect animal populations.

Of course, vaccination was not the only measure being taken to tackle COVID-19. As well as lockdowns and social distancing, people were encouraged to be fastidious in cleanliness. There was UK Government advice on washing hands – singing Happy Birthday whilst you did it, so you washed your hands for long enough – and posters such as shown in Figure 4.3 appeared in numerous places.

WASH
HANDS

DISINFECT
SURFACES

DISINFECT
HANDS

USE
FACE MASK

USE
RUBBER GLOVES

COVID-19 PREVENTION

Figure 4.3 A poster showing some of the preventative measures against COVID-19.
Sourced from Shutterstock curtesy of Volonoff, but made black and white.

Several therapies and remedies were also mooted. Some were bizarre and dangerous, whilst others were of dubious efficacy. Former US president Donald Trump suggested an "injection" of disinfectant, which would not be advisable.[15] There was a flurry of activity around the use of hydroxychloroquine. This was widely used as an anti-malaria drug, but it has also been used as an anti-inflammatory, for example, in rheumatoid arthritis (RA).[16] It has a wide range of actions in cells, including altering autophagy (cells self-eating), altering membrane stability and cell signalling processes. It was thought that some of its actions would help COVID-19 patients, but after several studies its efficacy was found not to be adequate. For example, it was tried in the laboratory with non-human primates,[17] specifically macaques, and the authors concluded that their "findings do not support the use of HCQ, either alone or in combination with AZTH [azithromycin], as an antiviral drug for the treatment of COVID-19 in humans." Eventually WHO stopped research on this drug for COVID-19.[18]

Other therapies centred around the use of anti-viral agents, such as remdesivir.[19] The use of some cytokine-based therapies was also suggested, as this will have direct effects on the inflammatory response. As death is often preceded by what is known as the "cytokine storm,"[20] the use of certain cytokines that can dampen this response makes sense. One treatment for COVID-19 that has not had much media attention, at least in the West, is the use of hydrogen gas. The use of molecular hydrogen (H_2) has been mooted for a range of human conditions, and even for the use in agriculture, but of pertinence here is its use as a COVID-19 remedy.[21] Patients would breathe the gas in for short periods of time and it has been reported to have significant effects. As far as we know this has never been used for a positive purpose on animals, but there is no reason why it should not be (assuming it can be practically administered). However, much of the research work has been carried out on animals. There has been a long history of using animals for research that may have benefits to humans. In the early days of looking at the effects of gases for medical use, such as oxygen, hydrogen, and nitrous oxide (laughing gas), animals were used, and even partially drowned to see the effects.[22] Animals are still used today for medical research, and the debate continues as whether this should carry on, especially for testing non-medical products (such as cosmetics) on animals,[23] with the subject even reaching the high courts.[24]

So what is known about the impact of SARS-CoV-2 variants and remedies in animals?

As mentioned, animals are often used for research experimentation and investigations around COVID-19 are no exception. However, we can learn much that can be translated (and licensed) for use on animals too, by such an approach (and it would be argued under a "One Medicine" approach that it is incumbent upon us to do so). Using a head dome, African green monkeys infected with SARS-CoV-2 were treated with inhaled remdesivir and the viral load was reduced,[25] indicating that this could be used as a treatment in non-human primates as well as humans. A similar remedy, i.e., favipiravir, but at high doses, was found to be effective in hamsters, whereas the treatment with hydroxychloroquine was not effective[26] (but as discussed, the idea of using hydroxychloroquine for COVID-19 was withdrawn). The authors hoped that this would lead to a better understanding of the use of such treatments and the generation of better alternatives in the future.

Any treatment deemed to be effective against SARS-CoV-2 in humans can, of course, be potentially beneficial for animals too. Of course, all animal species are not the same, and some drugs may work better in some species than others. However, in many cases, the basic biochemistry underpinning how an animal responds to a viral attack has similarities. Stress responses, such as a pathogen challenge, often lead to an increase in reactive oxygen species (ROS: such as superoxide anions and hydrogen peroxide) and reactive nitrogen species (RNS; such as nitric oxide: NO). This can lead to oxidative stress, or what some dub as nitro-oxidative stress,[27] and this leads to changes in the antioxidant responses. This cellular response is seen from nematodes to humans, and even in plants. In higher organisms, such as mammals, there is a response by the immune system, and, again, there are commonalities here. However, there are notable differences and, for this reason, some question the use of animals for medical research.[28] However, if we are considering a viral pathogen such as SARS-CoV-2, which we know can infect and even kill a range of animals, having robust data on what happens in those animals is useful, not only so we can cure humans, but also so that we can consider treatment for animals too. In that vein, some COVID-19 vaccines were developed for use in animals, and these will be considered next.

4.2 Vaccines developed in animals

Animals suffered in our fight to get a vaccine ready for the human population and to get a better understanding of the infection caused by SARS-CoV-2. Did any animals benefit from this research? In fact, vaccines were produced for animals and were used.

A cocktail of two antibodies against SARS-CoV-2 spike proteins and that prevent host cell invasions was found to be effective in both hamsters and rhesus macaques.[29] The authors conclude that the cocktail has therapeutic potential and would be entering clinical trials, but one assumes for human use. However, there is no reason why, if the mixture is effective, that it could not be used for animals deemed to be of high importance, for example non-human primates in zoos.

Others also took the antibody approach. Loo et al.[30] also used pairs of antibodies against the viral spike proteins and found positive effects in non-human primates – they also used macaques (cynomolgus macaques). Others too used these animals as well as mice to study the effects of antibodies.[31] Using only one antibody, LY-CoV555, Jones and colleagues[32] suggested that the data from macaques were robust enough to start clinical trials. Others looked at antibodies which were used against only very small regions (single domains) of proteins. The authors referred to these as synthetic nanobodies (or sybodies) and suggested that this was a good approach for vaccine production. Of relevance here is that this approach was tested in hamsters,[33] showing that non-primates may benefit too.

Worried about whether such approaches would be affected by viral variants, Bayarri-Olmos and colleagues concentrated on a virus variant from minks in which there was a change of one significant amino acid in the binding domain of the spike protein. This was a Y453F change, that is, there was a tyrosine (which as a hydroxyl group) at position 453 along the amino acid chain that was altered to be a phenylalanine, which also has a benzene ring as part of its structure, but no hydroxyl group, so a significant change. They found that the antibodies still worked.[34] Such data are reassuring in the search for therapies, but, of course, they do not cover all scenarios of possible amino acid changes.

Corbett and colleagues[35,36] concentrated on the use of mRNA-1273 in non-human primates, in particular three-year-old male and female rhesus macaques. They also carried out some experiments with Golden Syrian hamsters. They concluded that a two-dose regime was potentially beneficial against the virus.

Did such work lead to the development of a vaccine for use on animals?

On a quick look through the literature there appeared to be a study for a vaccine in gorillas,[37] but a deeper reading shows that this was a gorilla adenovirus-based system to be used in humans, and is, in fact, part of a clinical trial that seeks to recruit "Italian healthy volunteers aged 18–55 years and 65–85 years."[38] However, there was a call for projects focused on protecting animals.[39] And vaccines were used in animals, as outlined in the next section.

4.2.1 The use of vaccines on animals

With COVID-19 having such an anthropocentric focus, when discussing vaccines it is easy to find from the *Our World Data*[40] how many people have been vaccinated. As of 28 February 2023, the data stand at 13.31 billion doses given so far. Sadly, this still accounts for only 69.7% of the population, and only 27.7% of people in low-income countries have received at least one vaccine. However, these data are easy to find. How many animals have received a SARS-CoV-2 vaccination? Many animals get vaccinated (Figure 4.4) and it may surprise, but many have been vaccinated against SARS-CoV-2, and there are even choices of which vaccine to use.

In December 2021 it was reported that at San Diego Zoo and Safari Park approximately 260 animals had been vaccinated against SARS-CoV-2. These include a wide range of species.[41] Similarly Columbus Zoo and Aquarium vaccinated 110 animals, which were described as "high risk." Cheetahs were targeted for vaccination and the rollout included Bactrian camels, giraffes, as well as other animals.[42]

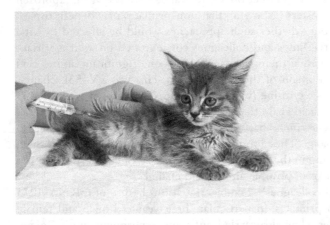

Figure 4.4 A kitten being injected as would be carried out if animals are vaccinated with the COVID-19 vaccine. Image from Shutterstock courtesy of Garna Zarina, but made black and white.

There are, in fact, several vaccines which have been made or are being developed for use in animals rather than humans. These are often based on the same antigen as used for the generation of human vaccines, but they include a different type of carrier protein for inducing a stronger immune response.

In December 2020 it was reported that the US Department of Agriculture (USDA) was saying that there was no need to license COVID-19 vaccines for pets, as the "data do not indicate that such a vaccine would have value."[43] But this did not stop such vaccines from being developed. One of the most widely used animal vaccines is produced by Zoetis.[44] This is described as a spike protein-based vaccine, and it showed good results in cats against the Delta variant (B.1.617.2) of the virus.[45] Zoetis has promised the delivery of 26,000 doses of its vaccine to be used in zoos and animal sanctuaries. These are not all in their home country either, and they are said to be spread across 13 nations. They are also hoping that the vaccine can be used in mink farms, again internationally. The vaccine was developed in dogs and cats, and, in their test animals, a good immune response was reported.

A second widely reported vaccine was developed in Russia.[46] This is the Carnivak-Cov vaccine (otherwise known as Karnivak-Kov).[47] It was reported to be the first which was animal specific, developed to protect carnivores, with immunity lasting approximately six months. It was developed on basis of an attenuated virus, inactivated either chemically, by heating or by radiation exposure. This vaccine was first developed using ferrets, as the company was hoping to target fur farms. Trials, completed in October 2020, were based on mink, arctic foxes, cats and rats, as well as some other animals. Data were reported that showed that it was working well and that it was safe, and currently the vaccine is being used in Russia as well as several other countries, including Greece, Austria and Poland.

A third vaccine is said to be under development by Applied DNA Sciences and Evvivax. In July 2022 these companies, along with several others, published a paper[48] reporting successful use of a linear DNA ("linDNA") vaccine encoding the binding domain of the viral spike protein in domestic cats.

Therefore, it can be seen that several companies in different countries are focusing efforts on the production and roll-out of vaccines against SARS-CoV-2 that can be used in animal populations. The design of such vaccines can be based on those used for the production of the human vaccines, and technology to produce the animal vaccines can also be translated across. However, none of these vaccines is particularly species specific; however, perhaps they do not need to be as the virus is the same, no matter the species it is infecting. As long as they give some protection and are safe, that is all that matters. It is likely that such vaccines will work in some animal species better than others, but any protection is better than none. If one has a precious pet, perhaps having it vaccinated will give peace of mind.

As mentioned in Chapter 3, vaccines have been used on black-footed ferrets (*Mustela nigripes*) with great success, although the article suggests that repeated inoculations will be required as the researchers state: "The vaccinated ferrets currently still maintain coronavirus antibodies. But those antibodies are showing signs of decreasing over time."[49]

4.3 Should animals be vaccinated when human populations are not?

The data given above show that a large number of people have yet to be vaccinated against COVID-19. As of the end of February 2023 there appears to be about 30% of the world's population still to be given at least one vaccination. Here in the United Kingdom, it is common for many people to have had four, or even five/six, jabs. And in the section above we discussed giving vaccinations to our pets.

This opens up an ethical dilemma. Should we even be considering vaccinating animals when humans are being ignored? Should we be using resources to make and distribute vaccines for use on animals when those resources could be used to get vaccines to people?

There are several considerations that we could debate:

■ Animals are often well loved and part of a wider family. People may say that their dog or cat is like a son or daughter to them. Are such animals not worth saving? COVID-19 did not have a devastating effect on the pet population, but the next pandemic might. Therefore, should we be offering salvation to pets as well as humans? On the other hand, whose pets should get the vaccine? Only those rich enough to afford it?

■ "Loved" animals may not only be domestic, but also some in captivity. Would a zoo keeper not wish to protect a gorilla that they have looked after and loved for years (equally for the visiting public)? It is interesting that Zoetis offered its vaccine to zoos and animal sanctuaries. This also has implications for animal conservation – indeed for some critically endangered animals (e.g., the Amur Leopard[50]), the zoo population greatly exceeds their population in the wild.

■ Animals are reservoirs for viruses. Should we not be stopping certain animal populations becoming "fermentation vessels" for viruses, and perhaps new variants? Vaccinating such animal communities may be of benefit to the animals and, in the longer term, humans too, as exemplified by the success of the rabies vaccination programmes in Germany, France, and other European countries.[51] On the other hand, not having animals caged and under poor welfare conditions in many instances may achieve the same outcome.

- Some animals are endangered. A vaccine one day might save the last of a species. Again, we have no knowledge of this happening during COVID-19, but it could happen next time. As discussed later, avian influenza is having a devastating effect across many bird communities. Should we not be prepared to do something about this, and would a vaccine be the answer?

Joel Baines, who is a professor of virology at Cornell University's Baker Institute for Animal Health, suggested that vaccines developed for human use should work on many animal species as that was the manner in which they were developed, but he goes further and suggests that there is no real need for animal vaccines, suggesting that animals either have mild symptoms or that they can be isolated from human contact. He also is quoted as saying, "However, these vaccines should be used in humans as a priority and it would be unethical to use a vaccine meant for humans to vaccinate an animal if vaccine doses are at all limiting."[52] This is obviously not the view of others, as animal vaccines have been created and will continue to be developed for use on animals. This will no doubt be perpetuated for any future pandemic/epidemic. After all, an outbreak in an animal population can spill over to humans wherever there is a close contact, be that companion animals or those on farms or in conservation sites.

As with any ethical debate, there is no correct conclusion. There are good reasons for the development of animal vaccines, and there are excellent reasons to reach human communities that are not being vaccinated. After all, such human communities would be reservoirs for viruses too, and places where variants may arise, and those variants will already be evolutionarily tailored to invade human cells. In an ideal world we would do both, but it would be easy for rich countries to take an "I'm now okay" attitude and perhaps be loath to pour finances into poorer countries to help in things such as vaccination programmes (after all, we have seen animal vaccines being used in rich countries, where humans remain unvaccinated in poorer ones). Despite some resources being given to vaccine rollouts,[53] some people such as the ex-PM of the United Kingdom, Gordon Brown, have been vocal about this issue, but with limited success.

As vaccines are continually developed for human use, whether it is against COVID-19 or a future virus, related vaccines will no doubt be produced and licensed to be used on animals, and for animal welfare, even on a relatively small scale, this is a good thing. It is most certainly in line with principles of reciprocity, where animals benefit from something that animals suffered and gave their lives in the development of (at a population not individual animal level, of course). The ethical debate will continue, but hopefully the plight of animals, whether loved or hated, whether big or small, will remain part of that debate.

4.4 Use of laboratory animal models

Animals have been vaccinated and perhaps saved from either a severe infection from SARS-CoV-2 or death. But to get those vaccines developed, whether for animal use or for rolling out across human populations, hundreds of experiments had to be carried out in numerous countries. Many experiments would be repeated, often in the race to be the first scientists to publish their data and get the glory. Often researchers do not know exactly what their competitors are doing, and they hope that they are either ahead of the competition or have a different slant on how to get to the aim of the work. Much human and animal work is carried out on cell lines, immortalised in laboratories so that experiments can be repeated quickly, efficiently, and on material that is well characterised. This is expensive, but it is often productive. However, to truly know what is happening under physiological conditions, animals are used.

Scientists have a list of "model organisms" that are used as testbeds for their work. Some of these are microbes, such as *Escherichia coli* (*E.coli*), a bacteria found in the gut. Plant scientists use a weedy plant: *Arabidopsis thaliana* (*A. thaliana*), which is part of the brassica family. It is small, easy to grow, and reproduces quickly. Animal scientists use nematode worms, such as *Caenorhabditis elegans* (*C. elegans*), which is again small, easy to grow, and no one much cares when it is killed. A common strain commonly used was isolated from a compost bin in Bristol,[54] near to where the authors work. They are common in gardens and are killed accidentally all the time. Very few ethics committees are concerned about the use of such organisms (although it should be noted that at the authors' institution, the Committee has set welfare standards, for example in relation to euthanasia). And because they are so well used, they are incredibly well characterised, and their genomes completely sequenced. They are a wonderful resource for scientists, with the pioneering work on this animal being carried out by Sydney Brenner.[55]

In 1986 the UK government passed the Animals (Scientific Procedures) Act, 1986, otherwise called ASPA. This restricts the use of vertebrates and cephalopoda in research in the United Kingdom, and to carry out such work permission must be sought from the Home Office. However, this obviously covers animals such as rats, mice, rabbits, guinea pigs, dogs, cats, and other animals that may be needed to evaluate COVID-19 vaccines.

As we have seen in the previous discussion, many vertebrates have been used in the development of COVID-19 vaccines, with macaques, hamsters, mice, dogs, and cats being discussed. These animals have suffered in efforts to protect the human population.

Gong and Bao[56] give an excellent summary of the animal models that were used in the research focused on SARS and MERS. These include primates such as the cynomolgus monkeys, African green monkeys, common marmosets, squirrel monkeys, moustached tamarins, and rhesus macaques, along with mice, hamsters, cats, rabbits, alpaca, goats, sheep, and horses.

For SARS-CoV-2, Takayama[57] also gives a useful list of animal models. Here there are no surprises, so the animals considered include mice, hamsters, ferrets, cats, and macaques. Gruber and colleagues helpfully compare what has been reported in hamsters, primates, and humans after SARS-Cov-2 infection,[58] which gives an indication how close animal models mimic the human disease. A comparison of what has been found with the use of model systems for infections caused by SARS-CoV, MERS-CoV, and SARS-CoV-2 was given by Natoli et al.[59] These authors had a particular focus on the effects seen in neuronal cells and the brain. The animals listed as being of use include non-human primates and mice.

A deeper look at the literature shows what such animal models are being used for in SARS-CoV-2 research. Under experimental conditions rabbits were shown to be able to be infected and symptoms included excretion of infectious virus from the throat and nose.[60] Similar results were reported in North American deer mice (*Peromyscus maniculatus*), although here the direct transmission between animals was also reported.[61]

Other researchers have used Syrian hamsters (*Mesocricetus auratus*). It was shown that the virus can replicate in the lungs of these animals, and symptoms reported to be similar to COVID-19 in humans,[62] although it has been suggested that the use of the Chinese hamster (*Cricetulus griseus*) is better,[63] partly as it is smaller but also because the symptoms seen on infection are greater. It was shown that the virus could be transmitted between individual hamsters, either via contact or as aerosols, and that "infection in golden hamsters resemble those found in humans with mild SARS-CoV-2 infections."[64] Syrian hamsters were used to test the efficacy of two neutralising antibodies as a possible therapy – they were shown to give a level of protection against infection.[65] Using the same animal model, it was shown that STATS signalling is involved in SARS-CoV-2 responses.[66] This mediated both lung injury and interferon responses. STATS proteins are involved in the expression of genes and hence protein production. They are induced to dimerise (come together in pairs, in this case in an antiparallel fashion), and, once in this form, they can bind to DNA and alter the action of transcription factors and, hence, expression of genes.

One of the characteristics of the COVID-19 pandemic was that it infected humans differently depending on the individual's age. Older people suffered more and were more likely to die, whereas many younger people seemed to be either asymptomatic or had much milder symptoms (although without doubt there were exceptions, with some apparently young, healthy people dying). In a study across a range of countries, including Italy, Canada, and several in the Far East (including China and Japan), Davies and colleagues stated, with regards to humans:

> We estimate that susceptibility to infection in individuals under 20 years of age is approximately half that of adults aged over 20 years, and that clinical symptoms manifest in 21% ... of infections in 10- to 19-year-olds, rising to 69% ... of infections in people aged over 70 years.[67]

[quote edited]

This age-related difference in infections and symptoms can be mimicked in animal models, as reported for the Syrian hamster.[68] Research with differently aged animals may therefore give an insight into what may be happening in humans and lead to benefits for patients, as well as social guidance, such as whether schools should be closed.

Primates have also been used. For example, the effects of remdesivir were studied in rhesus macaques, showing that the drug lowered viral load and reduced lung damage.[69] Alternative infection routes for the virus were also investigated in rhesus monkeys.[70] Both intranasal and intragastric transmission led to infection. Lu and colleagues[71] carried out a comparison of the infections in three primates. These were *Macaca mulatta* and *Macaca fascicularis,* representing the Old World monkeys, along with *Callithrix jacchus,* representing the New World monkeys. They concluded that "*M. mulatta* is the most susceptible to SARS-CoV-2 infection as compared to *M. fascicularis* and *C. jacchus.*" Using cynomolgus macaques (*Macaca fascicularis*) and rhesus macaques (*Macaca mulatta*) as model organisms, it was also shown that infection with SARS-CoV-2 leads to an alteration of gut microflora of these primates.[72]

Wild-type animal models are of use to scientists, but some researchers go further and genetically engineer the animals before testing to see if they could become infected. Using what was described as CRISPR/Cas9[73] knockin technology, the human ACE2 protein was introduced to mice and then infection investigated.[74] The authors report that such protein introduction increased viral load in these animals. The mice had worse symptoms than wild-type mice, but fatalities were not reported. Natoli et al.[75] also list the use of similar transgenic mice expressing the human ACE2 protein, saying that such animals can be infected with SARS-CoV, MERS-CoV, and SARS-CoV-2.

There has been some discussion about the secondary effects of our immune response. Did our own antibodies, produced to counter the SARS-CoV-2 virus, also attack our own cells, and hence lead to some of the symptoms reported? If so, did the same happen in animals and would this alter the effectiveness of a vaccine? Kanduc and Shoenfled[76] set about investigating this, and they concluded:

> only aged mice appear to be a correct animal model for testing an anti-SARS-CoV-2 spike glycoprotein vaccine to be used in humans.

No doubt animals will continue to be used in science, both to unravel some of the molecular biological mechanisms that underpin disease and to test potential drugs. However, the use of such animals has not been without criticism, with an excellent discussion being published by Sonia Shah in *The New Yorker* in February 2023.[77] She discusses why, in several cases, the effects seen in animals are not necessarily replicated in humans, and even suggests that the laboratory animals should be housed in different, more free range-like, conditions. But even with such changes in how experimental animals are cared for, they are still suffering in the name of advancing science, and the pandemic simply added to the use of animal models. Many of these animals were non-human primates, which we know have feelings and behaviours akin to humans.

To put this discussion about the use of animals into perspective, it is worth pausing over a paper by Philipp Schwedhelm and colleagues.[78] They asked the question: "How many animals are used for SARS-CoV-2 research?" It is an analysis focused on Germany but no doubt such numbers would be representative of many research-intensive countries (note: this study was published in October 2021, so not at the end of the pandemic). In Figure 1 of their paper they say:

> Between February 1, 2020 and July 27, 2021, 61,389 animals were approved for research projects related to SARS-CoV-2.

Although this seems like a colossal number, the authors then go on to say:

> In other words, since the outbreak of the pandemic in Germany, only 0.8% of all animals and 2.1% of projects using animals were authorized for research on SARS-CoV-2.

115

This data indicate the scale of the numbers that were used in the fight against COVD-19, but they also highlight the numbers of animals that must be used around the world for biological and biomedical research. In 2015 Katy Taylor and Laura Rego Alvarez[79] stated:

> We further extrapolated this estimate to obtain a more comprehensive final global figure for the number of animals used for scientific purposes in 2015, of 192.1 million.

This is a truly astonishing figure (although it is dwarfed by the >70 billion animals killed each year in the food industry[80]). It could be argued that animal use during the pandemic decreased as many laboratories stopped work as researchers could not travel and go to work. But it could be argued that many of those laboratory animals were still being housed, and perhaps many died without being used. Besides, such experiments were likely to have only been put on hold, and as research restarted the use of animals for investigations would have gone back to pre-pandemic levels, except COVID-19 research was being carried out too. It is a complex picture, but there is no doubt that thousands of animals suffered or had premature deaths because of SARS-CoV-2 research.

COVID-19 may be waning in some parts of the world, but even if it disappeared completely – which is unlikely – there is still the problem of long COVID, as discussed in Section 7.2. Animal models have been mooted for the study of this issue too,[81] so the use of animals for investigations into the effects of SARS-CoV-2 is not likely to end in the near future.

4.4.1 Use of animals for vaccine production

As well as vaccine containing the ingredients that give the specific immune response required, they need to be safe, and here animals have been used for testing (Figure 4.5).

Horseshoe crabs have been commonly used, although it is to be hoped that uptake of a synthetic alternative will increase. The blood of these crabs is blue, and animal welfare during the extraction process is regarded as very poor.[82] It was reported that along the east coast of the USA, companies "drained over 700,000 crabs in 2021."

The blue crab blood is particularly sensitive to bacterial toxins and has been used for many years been as a method for testing vaccine safety. In January 2021, an article from the National History Museum stated:

> If you have ever had a vaccine, chances are that it was tested for safety using horseshoe crab blood.[83]

Figure 4.5 Cartoon of a horseshoe crab donating its blood for vaccine use. Sourced from Shutterstock courtesy of Karazhanova Slava (image made black and white).

Therefore, the suffering of the horseshoe crab has ensured that human vaccines are not going to cause unwanted infections, and this would have continued during the COVID-19 pandemic, and as of July 2023 there had been 13,474,265,907 vaccine doses given.[84]

Further to this, some of the ingredients in vaccines are sourced from animals. As mentioned above, vaccines often contain adjuvants, which enable the immune system to mount a better response to the vaccine. One of the adjuvants commonly used is squalene from sharks (extracted from livers). According to *Shark Allies* even what was described as a "plant-based Covid-19 vaccine" uses a GSK product described as "'pandemic adjuvant' (which is MF59, made from shark squalene)."[85] A discussion of the safety and history of MF59 was given by Schultze and colleagues,[86] whilst a collection of articles on squalene use in vaccines and in cosmetics has been given by the *Shark Allies*.[87]

Therefore, as can be seen, there are unexpected and hidden involvements, and suffering, of animals in a pandemic, and each time we accept a COVID-19 vaccine we should be thankful for the animals enabling this to happen. The principle of reciprocity would suggest that where animals contribute to a vaccine/treatment, they should reciprocally benefit from these developments, where appropriate.

4.5 Vaccines of the future, will this transform their use in animals?

With vaccines being developed for human use, one question that should be asked is: will this improve development and rollout of vaccines for animals?

Vaccine development and use is not a trivial undertaking, and this is especially difficult in some locations of the world. For example, during the COVID-19 pandemic there was rapid development of vaccines that use mRNA technology, but, as pointed out, delivery of such vaccines may be difficult because: "ultracold chain currently unpractical for large-scale use in resource-limited settings."[88] Although such delivery difficulties may not be a problem for animals in captivity, such as at zoos, it will be much more difficult for animals in the wild, especially if at isolated locations.

During the COVID-19 pandemic there was some surprise, and indeed gratitude, that the vaccines were developed so quickly – from the announcement of the epidemic in China to the use of the first vaccine took a little under a year.[89] However, there is an ambition to be a lot quicker than this:

> The Coalition for Epidemic Preparedness Innovations' '100-day moonshot' aspires to launch a new vaccine within 100 days of pathogen identification.[90]

This is an ambitious aim, and one which will not be easy to achieve. However, as demand for rapid vaccines for humans increases, development and manufacture platforms will undoubtedly improve. There are several development platforms and vectors (examples of both listed by Excler et al. (2021)[91]) available for vaccine creation, all of which would be available for potential animal vaccines. These will no doubt be improved as time goes on, and such new technologies should, assuming the will and finances are in place, be adaptable to create vaccines tailored to animal use too, alongside the development of vaccines for human use. Ultimately it seems likely that any mass rollout of vaccines for animal use will be driven by either commercial (e.g., if a virus posed a serious risk to farm or companion animals) or conservation concerns, or, of course, concerns about the ways infected animals may affect human health. Vaccines for companion animals are more likely to be at the discretion of their owners and likely only available to those who can afford them.

Production capacity will, of course, be a likely constraining factor, as we saw with the COVID-19 pandemic (the UK Government cancelled a COVID-19 vaccine contract with Valneva, for example[92]). Richer nations will, of course, be taking measures to support future vaccine development, such as the UK Health Security Agency (UKHSA) Vaccine Development and Evaluation Centre (VDEC).[93,94] The UKHSA has published a report in August 2023 outlining how it is contributing to the "100 day moon-shot" referred to above, and this included manufacturing capability and capacity. This is reassuring to know, as is the statement that:

> The UK is approaching product design through a One Health per-spective with complementary research and development investments in human and animal health, such as in the Centre for Veterinary Vaccine Innovation and Manufacturing.[95]

This is an essential basis for tackling any emergent zoonotic disease. Time will tell, though, how far all these efforts accrue, in practice, within a con-text of "One Medicine," involving both human and veterinary clinicians, to the benefit of both humans and animals.

4.6 Conclusions

Several vaccines were rapidly developed for use on humans. It was rather amazing how quickly these were developed, tested, and then used at mass scale. Several companies and countries were involved, and millions of human lives were saved. A summary of the vaccines available is given in Table 4.3.

In August 2022, a website called *The Pig Site*[96] stated that at that time there had been 30 vaccines developed for human use but only three for animals. Interestingly they state that broad-range vaccines for animals are possible to develop:

> But it's challenging and expensive to develop and implement animal vaccines, and demand has been lacking as the broader health risk for animals isn't well known among the public. People tend to think only about their house pets.

Numerous animal species have been vaccinated, including gorillas, to try to protect them from a SARS-CoV-2 infection. In July 2021, it was reported that Zoetis had donated over 11,000 COVID-19 vaccine doses

Table 4.3 Examples of vaccines that had been developed against SARS-CoV-2

Vaccine name	Country	Approach used	Target species
Oxford/ AstraZeneca/AZD1222 (Covishield and Vaxzevria)	UK (Sweden): SII/ COVISHIELD produced in India	Adenovirus vector	Human
Moderna COVID-19 (mRNA 1273): Spikevax	USA	mRNA technology	Human
Pfizer/BioNTech Comirnaty	USA and Germany	mRNA technology	Human
Corbevax	USA	Spike protein	Human
CanSino Biologics: AD5-nCOV (Convidecia)	China	Viral vector	Human
Novavax: Nuvaxovid and Covovax	USA	Protein subunit	Human
Johnson & Johnson's Janssen/ Ad26.COV 2.S	Belgium	Viral vector	Human
Sinopharm COVID-19: BBIBP-CorV	China	Whole inactivated virus	Human
Sinovac-CoronaVac: PiCoVacc	China	Whole inactivated virus	Human
Bharat Biotech BBV152 COVAXIN	India	Whole inactivated virus	Human
Zoetis	USA based	Spike protein-based vaccine	Animals
Applied DNA Sciences and Evvivax	USA/Italy	Linear DNA ("linDNA") vaccine	Animals
Carnivak-Cov	Russia	Inactivated virus	Animals
Ancovax[104]	India	Inactivated SARS-CoV-2 (Delta) antigen	Animals

to help protect animals. This included to nearly 70 zoos. Across 27 states (USA) the vaccine has also be given to several academic institutions, as well as conservation sites and sanctuaries.[97] In June 2022 India announced its first animal vaccine against SARS-CoV-2, called Ancovax.[98]

Often such animals that are targeted for vaccination live in colonies and therefore such vaccination not only protects the individuals, but also the population they live with. Stopping such infections in these animals also reduces the chance of the virus mutating in a different animal species, with the risk that such viral variants can be re-transmitted back to the human population. New variants may be able to evade the immune system of a vaccinated individual, either animal or human, so for the continued efficacy of any vaccine being used it is important to limit the creation of new virus mutants. As can be seen from Table 3.2, there are a wide range of SARS-CoV-2 variants in the animal population already, and therefore limiting the development of any further ones is important.

Even though people working closely with animals, for example in zoos, were taking precautions, for example wearing PPE, animals still became infected. Therefore, vaccination of such animals seems a sensible thing to do. However, as discussed, there are ethical issues about using vaccines on animals when a significant proportion of the human population remains only partially vaccinated, or not vaccinated at all.

COVID-19 is not the only disease that is of concern for animals, of course, and other vaccines have been developed. An interesting one, for example, is one to be used on bees.[99] The vaccine was developed by Dalan Animal Health, Inc. and will hopefully protect bees from a disease called American Foulbrood disease, which is caused by *Paenibacillus larvae* (a Gram-positive, rod-shaped bacterium). It was given approval early in 2023 by the US Department of Agriculture (USDA).[100] The vaccine is mixed with the queen feed that is eaten by the workers. Subsequently it becomes incorporated into the royal jelly, which is then consumed by the queen. The subsequent larvae are also immune. The vaccine is based on killed whole-cell bacteria, and it is said to be organic.

Technological advances will also help with future vaccine developments. Recently, it has been shown that the use of artificial intelligence (AI) can enhance the design and therefore effectiveness of mRNA-based vaccines. A report in *Nature*[101] states that AI "yields jabs for COVID that have greater shelf stability and that trigger a larger antibody response in mice" and, hence, by extension can do the same for human vaccines. AI coding is only likely to improve in the future and therefore this is a significant advance for the development of future vaccines, for both humans and animals and for any future pandemic/epidemic, not just COVID-19.

As discussed later, other diseases may cause the next pandemic, and one of these is Avian Influenza. Vaccines for such diseases are also being developed.[102] Similarly, vaccines are being considered for monkeypox.[103] Vaccines will remain one of the main weapons in our armoury against many diseases, both in humans and in animals, as summarised in Table 4.3.

Notes

1 Robbins, S.C.C., Ward, K. and Skinner, S.R. (2011) School-based vaccination: A systematic review of process evaluations. *Vaccine, 29*, 9588–9599.

2 Bardell, D. (1996) Nestling cuckoos to vaccination: A commemoration of Edward Jenner. *BioScience, 46*, 866–871.

3 AJMC: A timeline of COVID-19 vaccine developments in 2021: https://www.ajmc.com/view/a-timeline-of-covid-19-vaccine-developments-in-2021 (Accessed 22/12/22).

4 Wang, L., Shi, W., Joyce, M.G., Modjarrad, K., Zhang, Y., Leung, K., Lees, C.R., Zhou, T., Yassine, H.M., Kanekiyo, M. and Yang, Z.Y. (2015) Evaluation of candidate vaccine approaches for MERS-CoV. *Nature Communications, 6*, 1–11.

5 Zhou, Y., Jiang, S. and Du, L. (2018) Prospects for a MERS-CoV spike vaccine. *Expert Review of Vaccines, 17*, 677–686.

6 Cid, R. and Bolívar, J. (2021) Platforms for production of protein-based vaccines: From classical to next-generation strategies. *Biomolecules, 11*, 1072.

7 CDC: Understanding how COVID-19 vaccines work: https://www.cdc.gov/coronavirus/2019-ncov/vaccines/different-vaccines/how-they-work.html (Accessed 28/02/23).

8 Excler, J.L., Saville, M., Berkley, S. and Kim, J.H. (2021) Vaccine development for emerging infectious diseases. *Nature Medicine, 27*, 591–600.

9 WHO: Coronavirus disease (COVID-19): Vaccines: https://www.who.int/news-room/questions-and-answers/item/coronavirus-disease-(covid-19)-vaccines?gclid=EAIaIQobChMIvoShzqK4_QIVlMftCh3QpgZ_EAAYAiAAEgK3g_D_BwE&topicsurvey=v8kj13 (Accessed 28/02/23).

10 Worldometer: United States Population (Live): https://www.worldometers.info/world-population/us-population/ (Accessed 22/12/22).

11 Guardian: Gordon Brown hits out at EU's 'neocolonial approach' to Covid vaccine supplies: https://www.theguardian.com/world/2021/aug/16/gordon-brown-hits-out-at-eu-neocolonial-approach-to-covid-vaccine-supplies (Accessed 22/12/220).

12 Institute of Government: Coronavirus vaccine rollout: https://www.institutefогgovernment.org.uk/article/explainer/coronavirus-vaccine-rollout (Accessed 02/06/23).

13 Our World in Data: SARS-CoV-2 sequences by variant: https://ourworldindata.org/grapher/covid-variants-bar?country=USA~GBR~ESP~ZAF~ITA~DEU~FRA~CAN~BEL~AUS (Accessed 28/02/23).

14 Cosar, B., Karagulleoglu, Z.Y., Unal, S., Ince, A.T., Uncuoglu, D.B., Tuncer, G., Kilinc, B.R., Ozkan, Y.E., Ozkoc, H.C., Demir, I.N. and Eker, A. (2022) SARS-CoV-2 mutations and their viral variants. *Cytokine & Growth Factor Reviews, 63*, 10–22.

15 NBC News: Trump suggests 'injection' of disinfectant to beat coronavirus and 'clean' the lungs: https://www.nbcnews.com/politics/donald-trump/trump-suggests-injection-disinfectant-beat-coronavirus-clean-lungs-n1191216 (Accessed 22/12/22).

16 Schrezenmeier, E. and Dörner, T. (2020) Mechanisms of action of hydroxychloroquine and chloroquine: Implications for rheumatology. *Nature Reviews Rheumatology, 16*, 155–166.

17 Maisonnasse, P., Guedj, J., Contreras, V., Behillil, S., Solas, C., Marlin, R., Naninck, T., Pizzorno, A., Lemaitre, J., Gonçalves, A. and Kahlaoui, N. (2020) Hydroxychloroquine use against SARS-CoV-2 infection in non-human primates. *Nature, 585,* 584–587.

18 Reuters: WHO halts trial of hydroxychloroquine in COVID-19 patients: https://www.reuters.com/article/us-health-coronavirus-who-hydroxychloroq/who-halts-trial-of-hydroxychloroquine-in-covid-19-patients-idUSKBN23O2T0? (Accessed 28/02/23).

19 Sheahan, T.P., Sims, A.C., Leist, S.R., Schäfer, A., Won, J., Brown, A.J., Montgomery, S.A., Hogg, A., Babusis, D., Clarke, M.O. and Spahn, J.E. (2020) Comparative therapeutic efficacy of remdesivir and combination lopinavir, ritonavir, and interferon beta against MERS-CoV. *Nature Communications, 11,* 1–14.

20 Montazersaheb, S., Hosseiniyan Khatibi, S.M., Hejazi, M.S., Tarhriz, V., Farjami, A., Ghasemian Sorbeni, F., Farahzadi, R. and Ghasemnejad, T. (2022) COVID-19 infection: An overview on cytokine storm and related interventions. *Virology Journal, 19,* 1–15.

21 Li, Y., Wang, Z., Lian, N., Wang, Y., Zheng, W. and Xie, K. (2021) Molecular hydrogen: A promising adjunctive strategy for the treatment of the COVID-19. *Frontiers in Medicine, 8,* 671215.

22 Hancock, J.T. and LeBaron, T.W. (2023) The early history of hydrogen and other gases in respiration and biological systems: Revisiting Beddoes, Cavallo, and Davy. *Oxygen, 3,* 102–119.

23 Grappe, C.G., Lombart, C., Louis, D. and Durif, F. (2021) "Not tested on animals": How consumers react to cruelty-free cosmetics proposed by manufacturers and retailers? *International Journal of Retail & Distribution Management, 49,* 1532–1553.

24 Guardian: High court rejects claim UK government 'secretly' ditched animal testing ban: https://www.theguardian.com/politics/2023/may/06/high-court-rejects-claim-uk-government-secretly-ditched-animal-testing-ban (Accessed 02/06/23).

25 Vermillion, M.S., Murakami, E., Ma, B., Pitts, J., Tomkinson, A., Rautiola, D., Babusis, D., Irshad, H., Seigel, D., Kim, C. and Zhao, X. (2021) Inhaled remdesivir reduces viral burden in a nonhuman primate model of SARS-CoV-2 infection. *Science Translational Medicine, 14,* eabl8282.

26 Kaptein, S.J., Jacobs, S., Langendries, L., Seldeslachts, L., Ter Horst, S., Liesenborghs, L., Hens, B., Vergote, V., Heylen, E., Barthelemy, K. and Maas, E. (2020) Favipiravir at high doses has potent antiviral activity in SARS-CoV-2–infected hamsters, whereas hydroxychloroquine lacks activity. *Proceedings of the National Academy of Sciences, 117,* 26955–26965.

27 Corpas, F.J. and Barroso, J.B. (2013) Nitro-oxidative stress vs oxidative or nitrosative stress in higher plants. *New Phytologist, 199,* 633–635.

28 Couzin-Frankel J. (2013) When mice mislead. *Science, 342,* 922–3, 925.

29 Baum, A., Ajithdoss, D., Copin, R., Zhou, A., Lanza, K., Negron, N., Ni, M., Wei, Y., Mohammadi, K., Musser, B. and Atwal, G.S. (2020) REGN-COV2 antibodies prevent and treat SARS-CoV-2 infection in rhesus macaques and hamsters. *Science, 370,* 1110–1115.

30 Loo, Y.M., McTamney, P.M., Arends, R.H., Abram, M.E., Aksyuk, A.A., Diallo, S., Flores, D.J., Kelly, E.J., Ren, K., Roque, R. and Rosenthal, K. (2022) The SARS-CoV-2 monoclonal antibody combination, AZD7442, is protective in nonhuman primates and has an extended half-life in humans. *Science Translational Medicine, 14,* eabl8124.

31 Li, D., Edwards, R.J., Manne, K., Martinez, D.R., Schäfer, A., Alam, S.M., Wiehe, K., Lu, X., Parks, R., Sutherland, L.L. and Oguin, T.H. (2021) The functions of

SARS-CoV-2 neutralizing and infection-enhancing antibodies *in vitro* and in mice and nonhuman primates. *BioRxiv*, 2020–12.

32 Jones, B.E., Brown-Augsburger, P.L., Corbett, K.S., Westendorf, K., Davies, J., Cujec, T.P., Wiethoff, C.M., Blackbourne, J.L., Heinz, B.A., Foster, D. and Higgs, R.E. (2021) The neutralizing antibody, LY-CoV555, protects against SARS-CoV-2 infection in nonhuman primates. *Science Translational Medicine, 13*, eabf1906.

33 Li, T., Cai, H., Yao, H., Zhou, B., Zhang, N., van Vlissingen, M.F., Kuiken, T., Han, W., GeurtsvanKessel, C.H., Gong, Y. and Zhao, Y. (2021) A synthetic nano-body targeting RBD protects hamsters from SARS-CoV-2 infection. *Nature Communications, 12*, 1–13.

34 Bayarri-Olmos, R., Rosbjerg, A., Johnsen, L.B., Helgstrand, C., Bak-Thomsen, T., Garred, P. and Skjoedt, M.O. (2021) The SARS-CoV-2 Y453F mink variant displays a pronounced increase in ACE-2 affinity but does not challenge antibody neutralization. *Journal of Biological Chemistry, 296.* DOI: https://doi.org/10.1016/j.jbc.2021.100536

35 Corbett, K.S., Werner, A.P., Connell, S.O., Gagne, M., Lai, L., Moliva, J.I., Flynn, B., Choi, A., Koch, M., Foulds, K.E. and Andrew, S.F. (2021) mRNA-1273 protects against SARS-CoV-2 beta infection in nonhuman primates. *Nature Immunology, 22*, 1306–1315.

36 Corbett, K.S., Flynn, B., Foulds, K.E., Francica, J.R., Boyoglu-Barnum, S., Werner, A.P., Flach, B., O'Connell, S., Bock, K.W., Minai, M. and Nagata, B.M. (2020) Evaluation of the mRNA-1273 vaccine against SARS-CoV-2 in nonhuman primates. *New England Journal of Medicine, 383*, 1544–1555.

37 Capone, S., Raggioli, A., Gentile, M., Battella, S., Lahm, A., Sommella, A., Contino, A.M., Urbanowicz, R.A., Scala, R., Barra, F. and Leuzzi, A. (2021) Immunogenicity of a new gorilla adenovirus vaccine candidate for COVID-19. *Molecular Therapy, 29*, 2412–2423.

38 ClinicalTrails.gov: Phase I study to assess the safety and immunology of a COVID-19 vaccine with GRAd-COV2 vaccine: https://clinicaltrials.gov/ct2/show/NCT04528641 (Accessed 28/02/23).

39 Sharun, K., Tiwari, R., Saied, A.A. and Dhama, K. (2021) SARS-CoV-2 vaccine for domestic and captive animals: An effort to counter COVID-19 pandemic at the human-animal interface. *Vaccine, 39*, 7119–7122.

40 Our World Data: Coronavirus (COVID-19) vaccinations: https://ourworldindata.org/covid-vaccinations (Accessed 28/02/23).

41 The Scientist: The rise of COVID-19 vaccines for animals: https://www.the-scientist.com/news-opinion/the-rise-of-covid-19-vaccines-for-animals-69503 (Accessed 28/02/23).

42 Columbus Zoo and Aquarium: Several species at the Columbus Zoo and Aquarium and the wilds receive COVID-19 vaccinations: https://www.colum-buszoo.org/news/several-species-columbus-zoo-and-aquarium-and-wilds-receive-covid-19-vaccinations (Accessed 16/07/23).

43 Science: Do we need a COVID-19 vaccine for pets? https://www.science.org/content/article/do-we-need-covid-19-vaccine-pets (Accessed 15/08/23).

44 Zoetis. (2021) Zoetis' emerging infectious disease capabilities support COVID-19 solutions for great apes and minks: https://www.zoetis.com/news-and-media/feature-stories/posts/zoetis-emerging-infectious-disease-capabilities-support-covid-19-solutions-for-great-apes-and-minks.aspx (Accessed 28/02/23).

45 Hoyte, A., Webster, M., Ameiss, K., Conlee, D.A., Hainer, N., Hutchinson, K., Burakova, Y., Dominowski, P.J., Baima, E.T., King, V.L. and Rosey, E.L. (2022)

Experimental veterinary SARS-CoV-2 vaccine cross neutralization of the Delta (B. 1.617. 2) variant virus in cats. *Veterinary Microbiology, 268,* 109395.

46 BBC: Covid: Russia starts vaccinating animals: https://www.bbc.co.uk/news/world-europe-57259961 (Accessed 28/02/23).

47 Chavda, V.P., Feehan, J. and Apostolopoulos, V. (2021) A veterinary vaccine for SARS-CoV-2: The first COVID-19 vaccine for animals. *Vaccines, 9,* 631.

48 Conforti, A., Marra, E., Palombo, F., Roscilli, G., Ravà, M., Fumagalli, V., Muzi, A., Maffei, M., Luberto, L., Lione, L. and Salvatori, E. (2022) COVID-eVax, an electroporated DNA vaccine candidate encoding the SARS-CoV-2 RBD, elicits protective responses in animal models. *Molecular Therapy, 30,* 311–326.

49 The Wildlife Society: Black-footed ferret COVID-19 vaccination seems to be working: https://wildlife.org/black-footed-ferret-covid-19-vaccination-seems-to-be-working/ (Accessed 21/07/23).

50 WWF: Amur Leopard: Probably the world's rarest cat? https://www.wwf.org.uk/learn/wildlife/amur-leopards (Accessed 02/06/23).

51 Müller, F.T. and Freuling, C.M. (2018) Rabies control in Europe: An overview of past, current and future strategies. *Revue scientifique et technique, 37*(2), 409–419. https://doi.org/10.20506/rst.37.2.2811.

52 The Pig Site: Why we don't have more COVID-19 vaccines for animals: https://www.thepigsite.com/articles/why-we-dont-have-more-covid-19-vaccines-for-animals (Accessed 16/05/23).

53 The Lancet: Challenges in the rollout of COVID-19 vaccines worldwide: https://www.thelancet.com/journals/lanres/article/PIIS2213-2600(21)00129-6/fulltext (Accessed 02/06/23).

54 University of Minnesota: Caenorhabditis Genetics Center (CGC): https://cgc.umn.edu/strain/N2 (Accessed 02/06/23) ["Isolated from mushroom compost near Bristol, England by L.N. Staniland"].

55 Goldstein, B. (2016) Sydney Brenner on the genetics of *Caenorhabditis elegans. Genetics, 204,* 1–2.

56 Gong, S.R. and Bao, L.L. (2018) The battle against SARS and MERS coronaviruses: Reservoirs and animal models. *Animal Models and Experimental Medicine, 1,* 125–133.

57 Takayama, K. (2020) In vitro and animal models for SARS-CoV-2 research. *Trends in Pharmacological Sciences, 41,* 513–517.

58 Gruber, A.D., Osterrieder, N., Bertzbach, L.D., Vladimirova, D., Greuel, S., Ihlow, J., Horst, D., Trimpert, J. and Dietert, K. (2020) Standardization of reporting criteria for lung pathology in SARS-CoV-2–infected hamsters: What matters?. *American Journal of Respiratory Cell and Molecular Biology, 63,* 856–859.

59 Natoli, S., Oliveira, V., Calabresi, P., Maia, L.F. and Pisani, A. (2020) Does SARS-Cov-2 invade the brain? Translational lessons from animal models. *European Journal of Neurology, 27,* 1764–1773.

60 Mykytyn, A.Z., Lamers, M.M., Okba, N.M., Breugem, T.I., Schipper, D., van den Doel, P.B., van Run, P., van Amerongen, G., de Waal, L., Koopmans, M.P. and Stittelaar, K.J. (2021) Susceptibility of rabbits to SARS-CoV-2. *Emerging Microbes & Infections, 10,* 1–7.

61 Griffin, B.D., Chan, M., Tailor, N., Mendoza, E.J., Leung, A., Warner, B.M., Duggan, A.T., Moffat, E., He, S., Garnett, L. and Tran, K.N. (2021) SARS-CoV-2 infection and transmission in the North American deer mouse. *Nature Communications, 12,* 3612.

62 Imai, M., Iwatsuki-Horimoto, K., Hatta, M., Loeber, S., Halfmann, P.J., Nakajima, N., Watanabe, T., Ujie, M., Takahashi, K., Ito, M. and Yamada, S. (2020) Syrian

hamsters as a small animal model for SARS-CoV-2 infection and countermeasure development. *Proceedings of the National Academy of Sciences, 117*, 16587–16595.

63   Bertzbach, L.D., Vladimirova, D., Dietert, K., Abdelgawad, A., Gruber, A.D., Osterrieder, N. and Trimpert, J. (2021) SARS-CoV-2 infection of Chinese hamsters (*Cricetulus griseus*) reproduces COVID-19 pneumonia in a well-established small animal model. *Transboundary and Emerging Diseases, 68*, 1075–1079.

64   Sia, S.F., Yan, L.M., Chin, A.W., Fung, K., Choy, K.T., Wong, A.Y., Kaewpreedee, P., Perera, R.A., Poon, L.L., Nicholls, J.M. and Peiris, M. (2020) Pathogenesis and transmission of SARS-CoV-2 in golden hamsters. *Nature, 583*, 834–838.

65   Rogers, T.F., Zhao, F., Huang, D., Beutler, N., Burns, A., He, W.T., Limbo, O., Smith, C., Song, G., Woehl, J. and Yang, L. (2020) Isolation of potent SARS-CoV-2 neutralizing antibodies and protection from disease in a small animal model. *Science, 369*, 956–963.

66   Boudewijns, R., Thibaut, H.J., Kaptein, S.J., Li, R., Vergote, V., Seldeslachts, L., Van Weyenbergh, J., De Keyzer, C., Bervoets, L., Sharma, S. and Liesenborghs, L. (2020) STAT2 signaling restricts viral dissemination but drives severe pneumonia in SARS-CoV-2 infected hamsters. *Nature Communications, 11*, 1–10.

67   Davies, N.G., Klepac, P., Liu, Y., Prem, K., Jit, M. and Eggo, R.M. (2020) Age-dependent effects in the transmission and control of COVID-19 epidemics. *Nature Medicine, 26*, 1205–1211.

68   Osterrieder, N., Bertzbach, L.D., Dietert, K., Abdelgawad, A., Vladimirova, D., Kunec, D., Hoffmann, D., Beer, M., Gruber, A.D. and Trimpert, J. (2020) Age-dependent progression of SARS-CoV-2 infection in Syrian hamsters. *Viruses, 12*, 779.

69   Williamson, B.N., Feldmann, F., Schwarz, B., Meade-White, K., Porter, D.P., Schulz, J., Van Doremalen, N., Leighton, I., Yinda, C.K., Pérez-Pérez, L. and Okumura, A. (2020) Clinical benefit of remdesivir in rhesus macaques infected with SARS-CoV-2. *Nature, 585*, 273–276.

70   Jiao, L., Li, H., Xu, J., Yang, M., Ma, C., Li, J., Zhao, S., Wang, H., Yang, Y., Yu, W. and Wang, J. (2021) The gastrointestinal tract is an alternative route for SARS-CoV-2 infection in a nonhuman primate model. *Gastroenterology, 160*, 1647–1661.

71   Lu, S., Zhao, Y., Yu, W., Yang, Y., Gao, J., Wang, J., Kuang, D., Yang, M., Yang, J., Ma, C. and Xu, J. (2020) Comparison of SARS-CoV-2 infections among 3 species of non-human primates. *BioRxiv*.

72   Sokol, H., Contreras, V., Maisonnasse, P., Desmons, A., Delache, B., Sencio, V., Machelart, A., Brisebarre, A., Humbert, L., Deryuter, L. and Gauliard, E. (2021) SARS-CoV-2 infection in nonhuman primates alters the composition and functional activity of the gut microbiota. *Gut Microbes, 13*, 1893113.

73   CRISPR/Cas9 is the technology being used for gene editing. For a review of this technology see: Yang, W., Yan, J., Zhuang, P., Ding, T., Chen, Y., Zhang, Y., Zhang, H. and Cui, W. (2022) Progress of delivery methods for CRISPR-Cas9. *Expert Opinion on Drug Delivery, 19*, 913–926.

74   Sun, S.H., Chen, Q., Gu, H.J., Yang, G., Wang, Y.X., Huang, X.Y., Liu, S.S., Zhang, N.N., Li, X.F., Xiong, R. and Guo, Y. (2020) A mouse model of SARS-CoV-2 infection and pathogenesis. *Cell Host & Microbe, 28*, 124–133.

75   Natoli et al. (2020).

76   Kanduc, D. and Shoenfeld, Y. (2020) Molecular mimicry between SARS-CoV-2 spike glycoprotein and mammalian proteomes: Implications for the vaccine. *Immunologic Research, 68*, 310–313.

77   The New Yorker: The case for free-range lab mice: https://www.newyorker.com/culture/annals-of-inquiry/the-case-for-free-range-lab-mice (Accessed 08/03/23).

78 Schwedhelm, P., Kusnick, J., Heinl, C., Schönfelder, G. and Bert, B. (2021) How many animals are used for SARS-CoV-2 research?: An overview on animal experimentation in pre-clinical and basic research. *EMBO Reports, 22,* e53751.

79 Taylor, K. and Alvarez, L.R. (2019) An estimate of the number of animals used for scientific purposes worldwide in 2015. *Alternatives to Laboratory Animals, 47,* 196–213.

80 Faunalytics: Global animal slaughter statistics & charts: 2022 update: https://faunalytics.org/global-animal-slaughter-statistics-charts-2022-update/ (Accessed 02/06/23).

81 Hofer, U. (2022) Animal model of long COVID? *Nature Reviews Microbiology, 20,* 446–446.

82 NPR: Coastal biomedical labs are bleeding more horseshoe crabs with little accountability: https://www.npr.org/2023/06/10/1180761446/coastal-biomedical-labs-are-bleeding-more-horseshoe-crabs-with-little-accountabi?ft=nprml&f =191676894 (Accessed 21/07/21).

83 Natural History Museum: Horseshoe crab blood: The miracle vaccine ingredient that's saved millions of lives: https://www.nhm.ac.uk/discover/horseshoe-crab -blood-miracle-vaccine-ingredient.html (Accessed 21/07/23).

84 WHO: WHO Coronavirus (COVID-19) dashboard: https://covid19.who.int/ (Accessed 21/07/23).

85 Shark Allies: Squalene Adjuvants in Vaccines: https://sharkallies.org/learn-about -shark-products/squalene-adjuvants-in-vaccines (Accessed 21/07/23).

86 Schultze, V., D'Agosto, V., Wack, A., Novicki, D., Zorn, J. and Hennig, R. (2008) Safety of MF59™ adjuvant. *Vaccine, 26,* 3209–3222.

87 Shark Allies: Squalene: Cosmetics + Vaccines: https://sharkallies.org/science-and -research#Squalene (Accessed 12/06/23).

88 Excler, J.L., Saville, M., Berkley, S. and Kim, J.H. (2021) Vaccine development for emerging infectious diseases. *Nature Medicine, 27,* 591–600.

89 National Institutes of Health: COVID-19 vaccine development: Behind the scenes: https://covid19.nih.gov/news-and-stories/vaccine-development (Accessed 09/03/23).

90 Joe, C.C., Segireddy, R.R., Oliveira, C., Berg, A., Li, Y., Doultsinos, D., Chopra, N., Scholze, S., Ahmad, A., Nestola, P. and Niemann, J. (2021) Accelerating manufacturing to enable large-scale supply of a new adenovirus-vectored vaccine within 100 days. *bioRxiv,* 2021–12.

91 Excler et al. (2021).

92 BBC: Scrapped Covid vaccine deal with Valneva cost UK taxpayers £358m: https://www.bbc.co.uk/news/uk-scotland-65949444 (Accessed 15/08/23).

93 UK Government: UKHSA unveils VDEC in 'step change' for UK's growing vaccine capabilities: https://www.gov.uk/government/news/ukhsa-unveils-vdec-in -step-change-for-uks-growing-vaccine-capabilities (Accessed 15/08/23).

94 UK Government: UKHSA's Vaccine Development and Evaluation Centre (VDEC): https://www.gov.uk/guidance/ukhsas-vaccine-development-and -evaluation-centre-vdec (Accessed 15/08/23).

95 UK Government: 100DM: How the UK is contributing to the global mission to develop pandemic-fighting tools within 100 days: 100DM - GOV.UK (www.gov .uk) (Accessed 15/08/23).

96 The Pig Site: Why we don't have more COVID-19 vaccines for animals: https://www.thepigsite.com/articles/why-we-dont-have-more-covid-19-vaccines-for -animals (Accessed 16/05/23).

97 Zoetis: Zoetis donates COVID-19 vaccines to help support the health of zoo animals: https://news.zoetis.com/press-releases/press-release-details/2021/Zoetis

-Donates-COVID-19-Vaccines-to-Help-Support-the-Health-of-Zoo-Animals
/default.aspx (Accessed 16/05/23).

98   The Hindu: Explained | Ancovax – India's first COVID-19 vaccine for animals:
     https://www.thehindu.com/sci-tech/science/explained-ancovax-indias-first
     -covid-19-vaccine-for-animals/article65516811.ece (Accessed 25/05/23).

99   Smithsonian Magazine: https://www.smithsonianmag.com/smart-news/the
     -worlds-first-vaccin e-for-honeybees-is-here-180981400/ (Accessed 16/01/23).

100 Businesswire: First-in-class honeybee vaccine receives conditional license from
     the USDA center for veterinary biologics: https://www.businesswire.com/
     news/home/20230104005262/en/First-in-Class-Honeybee-Vaccine-Receives
     -Conditional-License-from-the-USDA-Center-for-Veterinary-Biologics
     (Accessed 13/04/23).

101 Dolgin, E. (2023) 'Remarkable' AI tool designs mRNA vaccines that are more
     potent and stable. *Nature.* https://www.nature.com/articles/d41586-023-01487
     -y?utm_source=Nature+Briefing&utm_campaign=7222082a93-briefing-dy
     -20230503&utm_medium=email&utm_term=0_c9dfd39373-7222082a93-45
     565222 (Accessed 16/05/23).

102 Li, C., Bu, Z. and Chen, H. (2014) Avian influenza vaccines against H5N1 'bird
     flu'. *Trends in Biotechnology*, *32*, 147–156.

103 See, K.C. (2022) Vaccination for monkeypox virus infection in humans: A review
     of key considerations. *Vaccines*, *10*, 1342.

104 Developed by the Indian Council of Agricultural Research-National Research
     Centre on Equines (ICAR-NRCE).

5

Animal conservation and the pandemic

5.1 Introduction

Animals need to survive through a human disease pandemic, just as humans do. The previous discussion has concentrated on whether the susceptibility of animals to SARS-CoV-2 could be predicted, which animals were found to be directly affected by the virus, and whether we could mitigate the disease, either through treatment or through vaccines. However, there are a wide range of effects on animals that were indirect, and it is to those that we turn now.

Examples of the indirect effects of the pandemic on animals are to be found in the media (popular press and Internet) as well as in the science literature, but the true scale of the impacts will never be known. Here, several examples are given in the next two chapters, starting with the impact on animal conservation around the globe. However, this is far from a comprehensive treatise, but rather a range of examples on the topic.

5.2 The impact of the pandemic on animal conservation and zoos, wildlife parks, and aquaria

Animal conservation is a labour-intensive and often underfunded activity, but critical for the animals being considered. Conservation efforts are funded by a range of state and other actors, including conservation

DOI: 10.1201/9781003427254-5

organisations and other interested parties, amongst these being those facilities where animals are kept in captivity, for people to visit, for educational purposes, and to support conservation. As an example, the Association of Zoos and Aquariums (AZA) states that it supports "2500 conservation projects in more than 100 countries and spends on average \$160 million on conservation initiatives annually."[1] Therefore, this is not a trivial activity around the world. But what happened when the pandemic hit?

Besides direct donations, many conservation facilities rely on funding from people visiting. Ticket sales as people pass through the door are a lifeblood for many places. As countries went into lockdown, both local and international travel were severely restricted. People could no longer pop over to their local zoo with the kids for the day. Therefore, entry sales of tickets and in the related shops substantially plummeted.

Internationally, those places that rely on foreign income did not avoid the problem. International flights were curtailed, the tourist industry suffered, and this included conservation sites. Funding dwindled. Cristina Gomes, Assistant Director of the Tropical Conservation Institute (TCI), said on the Florida International University (FIU) web page:

"We know travel is going to be severely impacted," Gomes said. "This presents serious problems for conservation programs, especially in developing countries, that depend on tourism to support and fund their efforts."

Staff at some conservation sites would have been laid off, or furloughed, if such a scheme existed. Opportunities for training and internships would have been reduced or stopped, but the animals still needed to be cared for. As the *Los Angeles Times* stated in a headline[2]:

You can't furlough ferrets and tigers.

This highlights that although the public may not be coming to see the animals, the animals cannot simply be parked to wait for the pandemic to end. They still need feeding, cleaning, and veterinary care. Therefore, even if an organisation's income is down to nothing, their expenses do not dramatically drop. Some staff will still be required to look after the infrastructure of the site and to ensure that the animals' welfare is maintained. Some staff can be furloughed, such as those in the shop or on the ticket desks, but a core element of essential workers will have to be maintained.

The *Los Angeles Times* article went on to say that out of 240 zoos accredited by AZA, 90% of them closed during the pandemic. The Georgia Aquarium was estimated to be losing approximately \$2.5 million per month, a situation that is hard to maintain. As a measure to survive,

San Diego Zoo furloughed more than 1,500 employees. Zoos need to keep several months of revenue in reserve, but this was being quickly exhausted as the pandemic ran its course.

The situation for zoo and conservation organisations was further exacerbated by the difficulty of obtaining food for the animals and the price of supplies. The demand for animal feed remained stable – animals still needed to be fed - but supply chains were hit. Feed supplies, such as soybeans, flour, maize, barley, and wheat from Romania were reduced, as well as soybean from China because of either export bans or reduced action at ports, which affected India,[3] for example.

With expenses continuing and income curtailed, some conservation organisations were struggling (Figure 5.1).

Some newspaper headlines and articles tried to sensationalise the situation for some zoos and conservation sites. Locally (to the authors), Bristol Zoo Gardens was reported to be closing because of the pandemic, with headlines such as:

> Yet another lockdown victim: Bristol Zoo – the home of TV favourite Animal Magic - closes its doors after 186 years because of plunging visitor numbers caused by Covid restrictions.[4]

And:

> Bristol Zoo to close its doors after 186 years as animals are forced to move on.[5]

Animal Magic[6] was a children's TV show about animals fronted by a presenter called Johnny Morris, who was famous for making the animals

Figure 5.1 The entrance to Bristol Zoo Gardens in Clifton, Bristol. This site is now closed to the public, although as of the end of 2022 there are still animals present while the zoo is transitioned to its new site on the outskirts of the city. Photograph from Shutterstock, by urbanbuzz. Image has been made black and white.

pretend to talk with voice-overs. It ran from 1962 to 1984 on the BBC, and it helped to put Bristol Zoo on the map. The zoo was also famous for its Asian elephant, called Wendy.[7] Wendy, along with another (African) elephant called Christina, were even taken for walks in the local streets,[8] one assumes, in times where the prevailing ethical understandings about how (and if) large animals should be kept in zoos were "of its time," partly as a good advertisement for the zoo.[9]

The headlines about Bristol Zoo are interesting for two reasons. First, they make the situation sound dramatic and significant. Second, they were not really correct. Bristol Zoo may have been housed in a walled garden for a long time, and certainly it was a famous site in the Clifton area of the city. People used to talk, and complain, about the noise at night from some of the animals, but the zoo had been there long before the recent residents of the area. After the pandemic, Bristol Zoological Society,[10] which runs the zoo, did decide to close the Clifton site, but they had already created a new site on the outskirts of the city, in a place they named the Wild Place Project (recently renamed Bristol Zoo Project). The society is passionate about animal conservation, and even aids in teaching on this topic at the University of the West of England, Bristol (UWE). On their web page it says:

> Bristol Zoo Gardens plans to expand as they move to their Wild Place location in the next few years. This relocation will enable their team to develop future facing wildlife conservation projects and education facilities.[11]

This hardly gives the impression of the zoo "closing its doors." The old zoo site is earmarked for limited redevelopment, and being in a part of the city with one of the highest land values, it is likely that selling the site and moving did help to secure the zoo for the future, having survived through the pandemic. That is not to say that, in common with other zoos, they did not endure very tough times during the pandemic.

Bristol Zoo may have been able to survive, but there was a general call for governments to help such places with aid. On the BBC there were calls for the Welsh government to help:

> Frankie Hobro said her frustration was when the public recognised the plight of her Anglesey Sea Zoo and fundraised to save it, but the Welsh government failed to offer specific help.[12]

The zoo in Anglesey had expenses of £20,000 per month, meaning the closure was a real possibility. Others, for example at the Wales Ape and Monkey Sanctuary in Abercrave, echoed these views, calling for more support during difficult times. Animal welfare was the priority for such places until they had to decide to close or not. A small zoo in Devon, England, did close.[13] As the Zoological Society of London (ZSL: in *National Geographic*[14]) explained, their fixed monthly cost to feed and care for the animals was £1M. The DEFRA rescue package of £100M with a maximum individual award of £730,000 illustrates the harsh realities being experienced during this period and the lack of adequacy of government support to make any real dent in the problem.

Some animals were forced to move. It was reported that at Calgary Zoo the bamboo needed for the pandas could no longer be securely sourced due to flight problems, so the pandas were moved to China where they could be looked after.[15] In more troubling news, some zoos said that animals may have to be euthanised.[16] Statements made include, "German zoo may have to feed animals to each other," and a Neumünster Zoo spokesperson was reported to say, "We've listed the animals we'll have to slaughter first."

A spokesperson for the Hawk Conservancy Trust, UK,[17] where there is a large bird of prey collection and which relies for financing partly on admission fees to their site, said:

> We took a lot of strain as an organisation, but we have come out the other end, and in lots of ways would like to put the whole thing in the past.

Interestingly, there were also reports that the animals in some zoos were missing the human visitors, and appeared to be suffering emotionally. Seals, parrots, and pandas were listed as being particularly affected. Ms. Hachmeister from Berlin Zoo was quoted as saying: "The apes especially love to watch people." As staff were furloughed, animals were not always being cared for by their regular keepers, and this was also thought to be stressful to some animals.

Therefore, zoos, wildlife parks, and aquaria struggled in many ways, but they were desperately trying to keep animal welfare at the top of their agenda.

It is hard to estimate the financial impact of the pandemic on other conservation or animal welfare organisations, such as charities. A systematic look at this is beyond the scope of this book. The UK Charity Commission published survey findings (from charities across its remit) and stated, "Concerns around future viability remain. A third expect to generate less income from donations and fundraising in 2022,"[18] but this report also highlights the complex nature of the impacts. The data on this will also be emergent, and careful comparisons will need to be made between pre-pandemic and post-pandemic levels of funding. The geopolitical context has dramatically changed, not only in terms of economic impacts arising from the pandemic (and including here in the United Kingdom with a long history of "austerity" prior to the pandemic, and at the time of writing significant inflation and a major cost-of-living crisis) but also other serious impacts, such as the Russian invasion of the Ukraine. Whilst the pandemic caused its own issues for charities, it also contributed to conditions of national and international financial insecurity, which are continuing. It is clear that the charity sector remains under pressure. For example, in the United Kingdom, Barclays Bank has said of 2023:

> As 2023 unfolds, charities are in the middle of a perfect storm of growing demand, rising operating costs, and a widely predicted and significant income drop.[19]

Some are suggesting that a more diverse portfolio of fundraising will be necessary in the future. Charity Digital, for example, suggests a number of options for charities, including fundraising through corporate social responsibility, and it is easy to see how this might potentially be applied to conservation charities, if the appropriate businesses become engaged in this to fulfil their own synergistic corporate responsibility goals.

Gibbons et al.[20] published a view of the effects of the pandemic on biodiversity conservation globally. This highlights the very serious impacts we have already discussed, including a loss of tourism revenue (noting, e.g., that South Africa's national parks are about 85% funded by tourism-related spend, and that tourism revenue dropped by 90% between April and June 2020). They also highlight the potential for "reduced philanthropy" causing serious shortfalls, including:

> In late March 2020, 27% of environmental organizations in the United Kingdom reported they were either at high risk of becoming financially unviable in the coming months or had <4 months financial reserves remaining.
>
> (*Wildlife & Countryside Link*, 2020[21])

Mohan et al.,[22] in a UKRI funded project, found similarly concerning evidence of financial vulnerability of charities in England and Wales. It would be very interesting to understand how things "panned out" in practice for these organisations, and what the future will hold in an uncertain recovery context.

It is likely that donations to conservation charities will be impacted by the narratives that emerge in relation to causation, and these may equally apply to future major disease outbreaks or pandemics. For example, Shreedhar and Mourato[23] conducted an experiment to test three narratives, including origins amongst wild animals, a causal link with the human depletion of nature, and the possibility of blame on a biosecurity laboratory. It was found that the causal link with the human depletion of nature narrative and COVID-19 increased support for wildlife conservation policies. However, it had no significant effect on personal responsibility – they postulate that perhaps this is because whilst this narrative placed the responsibility on firms and government, the involvement of multiple actors may mean that people do not feel so personally responsible. They also found that:

> subjects increased donations after exposure to audio-visual narratives causally linking wildlife loss to human causes like poaching and habitat loss, compared to a control group omitting this causal information from the narrative.

They summarise their research thus:

> The results from this experiment suggest that there is scope to use this narrative to grow public engagement with extinction. This public support is key to craft a durable and legitimate long-term policy response to COVID-19, which concurrently addresses anthropogenic mass wildlife extinction.

This is just one exemplar of the multiplicity of factors that may impact the resources available to conservation and welfare charities in the coming years.

5.3 The impact on animals in the wild

It was not only animals in captivity that were at risk during the pandemic. In many places in the world, wild animals are cared for by those living in protected areas. It may be thought that because human movement was restricted during periods of the pandemic, animals may be safer. Humans could not transmit the virus to the wild animals if they were not there. On the other hand, close contact with some animals may be necessary for the protection of the animal populations. When this was the case, the increased use of PPE was recommended. Researchers were told to be careful, to approach wild animals only when strictly necessary. Any equipment being used on site had to be sterilised and then cleaned again afterwards. All PPE and other waste had to be removed from site and appropriately destroyed. Although it was not known if particular species could contract the virus from humans, it was not known that they could not, and, as discussed previously, predicting possible infections was not easy. Anyone in close contact with non-human primates or bats were viewed as most likely to cause a potential problem (and see, for example, the IUCN guidance on protecting bats from COVID-19[24]), but caution was a sensible approach in relation to all animals, especially mammals. Whether such a cautious approach was universally accepted is hard to tell, and one assumes that it probably was not in many places around the world and in many circumstances.

As of October 2022, it has been reported that there have been 675 natural outbreaks of the SARS-CoV-2 virus around the world which have affected a range of animal species, including humans and experimental work. In total, it is thought that 58 species of animal have been infected.[25] These, of course, are only the ones reported and known about. And, as discussed in Chapter 3, there is little known about many species. Some animals are often in close contact with humans, perhaps working for them, such as horses and elephants. Such species are also wild in parts of the world. Some species, such as sloths, are wild but are also in close contact with humans in sanctuaries, often visited by tourists. Although tourism based around animal watching and interactions was curtailed during lockdowns, as seen in the Surin province, Thailand,[26] this does not mean that such species in the wild are safe from SARS-CoV-2 or any future pandemic virus. At the present time, molecular biology has limited use here too, as often such species as sloths and elephants are not represented in the genomic data available. Testing of wild animals is also difficult, although some testing of elephants in India in the wake of the death of a lion was carried out[27] (although this did not affect the increased poaching that was reported too[28]). However, the full extent of the infections of many species will never be known (Figure 5.2).

Figure 5.2 A three-toed sloth. Photograph from Shutterstock, taken by Sofie Hojabri. Image has been made black and white.

However, it was not only the direct effect of the virus that was a potential issue. In some areas of the world, human movement was restricted, and this meant that there was a reduction in security of some animal species. This resulted in more animal poaching reported. Behera et al.[29] said that "illegal entry into forests and illegal hunting of ungulates and small animals increased significantly during this lockdown period." Looking at this type of activity in southern Africa, Brian Lucas[30] said that poaching as a supply of animal products for international trafficking was reduced, probably as travel was curtailed and borders closed. On the other hand, poaching for sustenance and consumption, for example, bushmeat, was generally increased. It was interesting that he also suggested that the pandemic made the collection of robust data difficult, partly because of funding reductions. Perhaps this is something that requires more support in the future when we have another epidemic.

A similar rise in poaching was reported in Nepal,[31] and here it was commented that animals were able to roam more because of the reduction of human activity. The trading in pangolins was also increased in India,[32] and, rather worryingly, the authors say that the trade is probably much greater than they could determine. On the other hand, the poaching of rhinoceros was said to be "paused" in South Africa (at Kruger National Park) because of COVID-19 but increased again as the pandemic waned.[33] The temporary decrease in poaching of these animals may have been due to the difficulty of trading the horns across international borders during lockdown. However, in contrast, there have been some reports of an increase in animal deaths associated with fear of the virus. For example, the David Shepherd Wildlife Foundation noted in an article written in 2020:

> Another worrying trend occurring concerns local superstitions in southern Africa. Rangers have been finding mutilated lion corpses, missing heads, tails, and paws. Known as muti, these body parts are used in traditional medicine because of their association with prosperity and good luck, suggesting that local peoples in fear of the coronavirus are seeking alternative medicine from wildlife.[34]

As pointed out by Guynup et al. (2020),[35] the Convention on International Trade in Endangered Species of Wild Fauna and Flora (CITES) regulates the international trade in plants and animals. This was an agreement signed by 183 nations and covers at least 35,000 species. But these authors also say that poaching, trafficking, and illegal sales of such species occur in every country in the world. Traditional Chinese medicines alone involve around 1,500 species, so with such a big global trade in animals, it is unlikely to be stopped in the near future.

Less human movement leads to more animal movement, which can lead to proximity issues with the local human population. With poaching increased, there were also "accidental" killings of animals. For example, in Uganda a poacher killed a gorilla "in self defence" after the gorilla charged at him.[36] The poacher was using spears to kill a bush pig, but the gorilla was close by. The report claims that the killing was "a direct result of COVID-19," probably because of the lack of tourism during lockdown and therefore less human movement influencing the gorillas' movements. As a deterrent to others, the poacher was given an 11-year prison sentence.

In India, there was a substantial increase in the ingress of animals into urban areas, and there was a large increase in human/animal conflicts.[37] Animals listed included tigers, leopards, and elephants. In a rather alarming table, a list of animal deaths is given, including a king cobra, elephants, and bears. One of the elephants was pregnant. However, a further reading of the text and table gives a better context behind these animal deaths, as it is stated that "wild animals including leopards, tigers, and elephants have killed as many as 13 people in different parts of the State, Madhya Pradesh." Animals may assume that the quiet of lockdown is an opportunity to roam into urban space, or that the closure of the wildlife parks may have made tracking of individual animals by rangers less easy, but that does not mean that the humans are no longer there. The interaction of wild animals, often large and dangerous, with human communities is fraught with conflicting interests, and unfortunately, both sides suffer. There is much to be learnt here for the future. Clearly, managing how animals move into and out of urban or marginal space needs to be considered for future lockdowns, and perhaps further research and resources should address safe (high welfare), species-specific, deterrent measures.

As further discussed below, closer to home, it was widely reported that wild goats were freely roaming through the streets of the Welsh seaside town, Llandudno, in March 2020.[38] This was seen as a curiosity and a minor nuisance at worst, as the goats themselves were not harmful to the human population and rather attractive, charismatic animals. It is striking that this "goat incursion" was reported on 30 March 2020, only one week into the full UK lockdown, which many commentators saw as a sign of the resilience of nature to reclaim lost territory once humans are out of the picture. Such effects have been seen at many abandoned sites worldwide, most notably the Chernobyl Exclusion Zone.[39] These examples raise the hope that the damage humans are currently inflicting on the natural environment is not yet irreversible, and that once we are gone, nature will quickly reclaim the areas we have laid waste to.

Susan Lappan and her colleagues[40] see the end of the pandemic as an opportunity to change the behaviour of humans, and not to go back to "business as usual." They concentrate on the human–primate interface and suggest that there is now the chance to reduce aggression between primates and human communities, to reduce the keeping of primates as pets, and to reduce the potential for future zoonosis. They suggest that primatologists pave the way for others to follow in the progression of future conservation. Perhaps the pandemic will have long-term positive consequences for some animals in the wild.

An interesting and rather unexpected effect of the COVID-19 pandemic on animals was the change that was noted in birdsong. Derryberry and colleagues,[41] publishing in *Science,* stated:

> We show that noise levels in urban areas were substantially lower during the shutdown, characteristic of traffic in the mid-1950s. We also show that birds responded by producing higher performance songs at lower amplitudes, effectively maximizing communication distance and salience.

Here was a relatively rapid change in animal behaviour, and one that will no doubt revert back to how it was before as soon as humans resume their normal behaviour.

5.4 Conclusions

Organisations that aim to have an impact in animal conservation suffered during the COVID-19 pandemic. Restricted human movements have a range of effects, from limiting income from tourism to reduced security of vulnerable animal populations. Food security was an issue, with both accessibility and price increases having an impact. Some animals had to be moved, and euthanasia programmes were mooted.

With animal welfare high on the agenda for conservation groups, many people had to be laid off or furloughed. The human cost was high in this industry, as it was in numerous others, especially the tourist industry in general.[42]

It appears that animals suffered in surprising ways too. Animals in zoos were said to miss the humans, which they appeared to like (or at least be used to) watching. Without the human activity and noise, zoos must have been strange places, and it seems as though the animals noticed.

There was a lot of negative reporting of the situation in conservation groups, including zoos, during the pandemic, with some alarming headlines. Now, as we hopefully are approaching the tail end of the pandemic, most of the organisations have survived and are building back a future. Many will have exhausted financial reserves, reined back on investment projects, sold land, and, in some cases, relied on donations to keep themselves going, but for the foreseeable future they have a way to continue and build back their valuable work in conservation, education, and maintaining animal welfare.

Animals in the wild suffered too. Poaching in most cases was increased, and although international trafficking of animal products may have been temporarily curtailed, this was not a lasting phenomenon. Animal–human conflicts increased in some parts of the world as animals roamed closer to communities, which feared for the safety and security of their crops.

The future may be brighter if lessons are learnt. Security of vulnerable animal communities needs to be maintained, and we have a lot to learn about embracing the roaming of animals into our human communities, or this needs to be managed more sustainably. As discussed later, humans need to have a more One World view and live with our environment rather than continually trying to tame it. Having a more respectful opinion of our animal communities will be good for them, and also good for humans too, perhaps decreasing the potential for future zoonotic diseases.

Notes

1 Association of Zoos and Aquariums (AZA): Conservation funding: https://www.aza.org/conservation-funding/ (Accessed 22/12/22).
2 Los Angeles Times: Editorial: You can't furlough ferrets and tigers. Zoos need federal funds during the pandemic, too: https://www.latimes.com/opinion/story/2020-06-25/zoos-need-federal-funds-during-the-pandemic-too (Accessed 22/12/22).
3 All About Feed: Covid-19: The impact on the animal feed industry: https://www.allaboutfeed.net/animal-feed/raw-materials/covid-19-the-impact-on-the-animal-feed-industry/ (Accessed 22/12/22/).
4 Mail Online: Yet another lockdown victim: Bristol Zoo: https://www.dailymail.co.uk/news/article-11171673/Bristol-Zoo-closes-doors-186-years-plunging-visitor-numbers-caused-Covid.html (Accessed 23/12/22).
5 Express: Bristol Zoo to close its doors after 186 years as animals forced to move on: https://www.express.co.uk/news/nature/1566912/Bristol-zoo-closing-down-date-pandemic-lockdown (Accessed 23/12/22).
6 IMDb: Animal magic: https://www.imdb.com/title/tt0201375/ (Accessed 23/12/22).
7 Bristol Live: Never forget Wendy the Bristol Zoo elephant who became a TV star: https://www.bristolpost.co.uk/news/history/gallery/never-forget-wendy-bristol-zoo-7530205 (Accessed 23/12/22).
8 BBC: Wendy's life in picture: https://www.bbc.co.uk/bristol/content/news/2002/09/12/news2/gallery.shtml (Accessed 23/12/22).
9 Bristol Zoo's elephant called Wendy died at the age of 42. An article about it, from the BBC and updated in 2002, can be found here: https://www.bbc.co.uk/bristol/content/news/2002/09/12/news2/wendy.shtml (Accessed 23/12/22).

10 Bristol Zoological Society: About us: https://future.bristolzoo.org.uk/about-us/ (Accessed 23/12/22).

11 UWE: Wildlife ecology and conservation science: https://courses.uwe.ac.uk/45MN/wildlife-ecology-and-con (Accessed 23/12/22).

12 BBC: Coronavirus: 'Utterly irresponsible' attitude to zoos: https://www.bbc.co.uk/news/uk-wales-53050225 (Accessed 23/12/22).

13 The Print: What is the future of zoos after Covid? We can't afford them going extinct: https://theprint.in/environment/what-is-the-future-of-zoos-after-covid-we-cant-afford-them-going-extinct/490956/ (Accessed 23/12/22).

14 National Geographic: Zoo crisis deepens amidst second national lockdown and 'restrictive' bailout conditions: https://www.nationalgeographic.co.uk/animals/2020/11/zoo-crisis-deepens-amidst-second-national-lockdown-and-restrictive-bailout (Accessed 02/06/23).

15 Calgary Zoo: If you can't bring bamboo to the giant pandas, you need to move the giant pandas closer to the bamboo!: https://www.calgaryzoo.com/GiantPandas (Accessed 23/12/22).

16 BBC: Coronavirus: German zoo may have to feed animals to each other: https://www.bbc.co.uk/news/world-europe-52283658 (Accessed 23/12/22).

17 Hawk Conversancy Trust: https://www.hawk-conservancy.org/ (Accessed 26/07/23).

18 Charity Commission: Concerns around future viability: https://www.gov.uk/government/publications/charity-commission-covid-19-survey-2021/covid-19-survey-2021#future-outlook (Accessed 02/06/23).

19 Barclays: Charities' resilience will be tested this year: https://www.barclayscorporate.com/insights/industry-expertise/2023-outlook/charities/ (Accessed 16/07/23).

20 Gibbons, D.W., Sandbrook, C., Sutherland, W.J., Akter, R., Bradbury, R., Broad, S., Clements, A., Crick, H.Q., Elliott, J., Gyeltshen, N. and Heath, M. (2022) The relative importance of COVID-19 pandemic impacts on biodiversity conservation globally. *Conservation Biology*, *36*, e13781.

21 Heritage Fund: Environment and Conservation Organisations Coronavirus Impact Survey Report: https://www.heritagefund.org.uk/sites/default/files/media/attachments/Coronavirus%20eNGO%20survey%20analysis%20report_1.pdf (Accessed 02/06/23).

22 University of Birmingham: Financial vulnerability in UK charities under Covid-19: An overview: https://www.birmingham.ac.uk/documents/college-social-sciences/social-policy/tsrc/financial-vulnerability-in-uk-charities-under-covid-19.pdf (Accessed 02/06/23).

23 Shreedhar, G. and Mourato, S. (2020) Linking human destruction of nature to COVID-19 increases support for wildlife conservation policies. *Environmental and Resource Economics*, *76*, 963–999.

24 IUCN: Protecting bats from COVID-19: https://www.iucnbsg.org/bsg-publications.html (Accessed 02/06/23).

25 Agnelli, S. and Capua, I. (2022) Pandemic or panzootic—A reflection on terminology for SARS-CoV-2 infection. *Emerging Infectious Diseases*, *28*, 2552–2555.

26 Kungwansupaphan, C. (2021) The socio-economic impact of COVID-19 on Khunchaitong Elephant community-based tourism in Surin province, Thailand. *Journal of Mekong Societies*, *17*, 28–49.

27 Phys.org: Elephants in India tested for coronavirus after rare lion's death: https://phys.org/news/2021-06-elephants-india-coronavirus-rare-lion

.html#:~:text=Twenty%2Deight%20elephants%20have%20been,Asiatic %20lion%20from%20the%20virus (Accessed 15/08/23).

28 Independent: Covid escalates elephant killings in eastern India to 'crisis proportions': https://www.independent.co.uk/asia/south-asia/elephant-deaths -covid-india-poaching-b1876632.html (Accessed 15/08/23).

29 Behera, A.K., Kumar, P.R., Priya, M.M., Ramesh, T. and Kalle, R. (2022) The impacts of COVID-19 lockdown on wildlife in Deccan Plateau, India. *Science of The Total Environment, 822,* 153268.

30 Lucas, B. (2022) Impact of COVID-19 on poaching and illegal wildlife trafficking trends in Southern Africa: https://opendocs.ids.ac.uk/opendocs/ bitstream/handle/20.500.12413/17179/1094_Impact_of_COVID-19_on _poaching_and_illegal_wildlife_trafficking.pdf (Accessed 19/02/23).

31 Koju, N.P., Kandel, R.C., Acharya, H.B., Dhakal, B.K. and Bhuju, D.R. (2021) COVID-19 lockdown frees wildlife to roam but increases poaching threats in Nepal. *Ecology and Evolution, 11,* 9198–9205.

32 Aditya, V., Goswami, R., Mendis, A. and Roopa, R. (2021) Scale of the issue: Mapping the impact of the COVID-19 lockdown on pangolin trade across India. *Biological Conservation, 257,* 109136.

33 Ferreira, S.M., Greaver, C., Simms, C. and Dziba, L. (2021) The impact of COVID-19 government responses on rhinoceroses in Kruger National Park. *African Journal of Wildlife Research, 51,* 100–110.

34 David Shepherd Wildlife Foundation: Sparking change: What we can learn from the Covid-19 pandemic: https://davidshepherd.org/news/sparking -change-what-we-learn-from-covid-19/ (Accessed 02/06/23).

35 Guynup, S., Shepherd, C.R. and Shepherd, L. (2020) The true costs of wildlife trafficking. *Georgetown Journal of International Affairs, 21,* 28–37.

36 Kalema-Zikusoka, G., Rubanga, S., Ngabirano, A. and Zikusoka, L. (2021) Mitigating impacts of the COVID-19 pandemic on gorilla conservation: Lessons from Bwindi Impenetrable Forest, Uganda. *Frontiers in Public Health,* 920.

37 Sarkar, C., Hazarik, S.S. and Singh, B. (2020) The impact of COVID-19 lockdown period on man and wildlife conflict in India. *International Journal of Advanced Research in Engineering and Technology (IJARET), 11*(9). https:// doi.org/10.34218/IJARET.11.9.2020.063.

38 BBC: Coronavirus: Goats take over empty streets of seaside town: https:// www.bbc.co.uk/news/uk-wales-52103967 (Accessed 02/06/23).

39 UN: How Chernobyl has become an unexpected haven for wildlife: https:// www.unep.org/news-and-stories/story/how-chernobyl-has-become-unex- pected-haven-wildlife (Accessed 02/06/23).

40 Lappan, S., Malaivijitnond, S., Radhakrishna, S., Riley, E.P. and Ruppert, N. (2020) The human–primate interface in the New Normal: Challenges and opportunities for primatologists in the COVID-19 era and beyond. *American Journal of Primatology, 82,* e23176.

41 Derryberry, E.P., Phillips, J.N., Derryberry, G.E., Blum, M.J. and Luther, D. (2020) Singing in a silent spring: Birds respond to a half-century soundscape reversion during the COVID-19 shutdown. *Science, 370,* 575–579.

42 Gössling, S., Scott, D. and Hall, C.M. (2020) Pandemics, tourism and global change: A rapid assessment of COVID-19. *Journal of Sustainable Tourism, 29,* 1–20.

The indirect effects of COVID-19 on animals

6.1 Introduction

COVID-19 had many effects on animals, from direct infection to conservation groups struggling to maintain their valuable work. On top of these impacts on animals, there were many effects which could be considered to be of an indirect consequence of the SARS-CoV-2 outbreaks. It is to these that we turn now.

Many of the indirect effects of the pandemic were almost unnoticed, particularly by the popular media. However, significant animal suffering resulted. And, like so many aspects of the story about animals during the pandemic, the full extent of the issue will probably never be known. Some of the effects were a direct consequence of human action, such as culling programmes. Were these needed? This could be debated, but now they have happened, it is too late for those particular individual animals. Some of the effects were unconscious human action. It is easy to discard your waste and not worry about where it goes and what happens to it. However, living in a complex ecosystem means there are consequences, and often it is the animals around us that suffer.

This chapter will endeavour to capture some of the issues which arose via the indirect consequences of the SARS-COV-2 pandemic.

6.2 Human perception of animals as transmitters

The SARS-CoV-2 virus almost certainly emanated from a bat population in China. It is well known that bats harbour a host of coronaviruses. In a study of coronavirus and bats in Costa Rica, the authors say:

DOI: 10.1201/9781003427254-6

The finding of diverse CoV sequences in one specific bat species emphasizes that the whole diversity of CoV is still unknown and that surveillance should be continued.[1]

In another study from Costa Rica, coronavirus was found in both nectivorous (e.g. *Glossophaga soricina*) and frugivorous (e.g. *Artibeus jamaicensis, Carollia perspicillata, Carollia castanea*) bats.[2] In a similar study from Korea,[3] betacoronaviruses were found associated with the bat *Rhinolophus ferrumequinum*, although the authors stressed that this was not related to SARS-CoV-2. Alphacoronaviruses were also detected in association with bats, with a range of species investigated, including *Hypsugo alaschanicus, Miniopterus fuliginosus, Miniopterus schreibersii, Rhinolophus ferrumequinum, Myotis bombinus, Myotis macrodactylus,* and *Myotis petax*. The authors advocated continued monitoring of such species, which sounds like good advice. Papers were published long before the COVID-19 pandemic (e.g., Smith and Wang, 2013[4]) voicing concerns that bats were the potential source of future SARS-like viruses, but also others, including the Nipah virus, Hendra virus, and even the Ebola virus. It is easy to see why bats were getting a bad reputation.

SARS probably emanated from the bats (possibly genus *Rhinolophus*[5]), MERS from dromedary camels, and COVID-19 probably came from the horseshoe bat, *Rhinolophus affinis*. Therefore, animal species were the likely source of all these epidemic coronaviruses.

It was early in the COVID-19 pandemic that it was announced that bats were the source of the virus (as already mentioned, most likely from *Rhinolophus affinis*). With the notion, which became embedded in the "facts" being bandied about, that SARS-CoV-2 was the "fault" of a bat, and millions of humans suffering, meant that bats became the enemy. Bats often had a bad reputation anyway, with their activity being associated with vampires, and spooky places. Vampire bats are also known to target cattle, especially in South and Central America, where they can be a problem for farmers.[6] There is no doubt that vampire bats can look a little menacing, as seen in Figure 6.1, and they are a major concern regarding rabies virus infection.[7] Therefore, many people either fear or hate bats, or both. There are even Internet forums about bats that include questions such as if they will attack humans while flying.[8]

Therefore, when the pandemic came, and bats were blamed, their reputation suffered even more. There were reports that bats were subsequently being persecuted. In China, bats were animals traditionally associated with happiness and good luck, but calls were made to protect bats from being forced from hibernation in cities,[9] which, of course, would severely decrease their well-being and likelihood of survival. Some have called for mass culling of bats in an effort to protect human populations.[10]

Figure 6.1 Common vampire bats (*Desmodus rotundus*). Photograph from
Shutterstock, courtesy of belizar. Image has been made black and white.

Unfortunately, there seemed to be little discrimination of bat species,
with all bats being branded as bad. Clearly, many bat species do harbour
viruses that potentially can "jump" to humans, but destroying bat populations is not the way to control such threats.

It is worth mentioning here that there has been a deliberate effort
in recent years by the WHO to avoid naming viruses in a manner that
implicates specific geographical locations or species. This is partially in
reaction to the mass culling of pigs, which took place in Egypt during
the "Swine Flu" pandemic of 2009,[11] a move that would have had no
effect on viral transmission as by this point the virus was spreading mainly
human to human.

However, it was not thought that bats transmitted the SARS-CoV-2
virus to humans directly. Despite the arguments as to whether the virus
escaped from a laboratory in Wuhan or whether it followed another route
for transmission from a bat to humans,[12] several animals were singled out.
One of these was the pangolin, with reports appearing in February 2020
that they were involved.[13] Suddenly, pangolins were also seen as bad. The
poor creatures are endangered anyway, being dubbed the most trafficked
mammal, and the negative vibe caused by SARS-CoV-2 did not help
with their welfare.[14] Even posters, as seen in Figure 6.2, seem to give a
clear indication that the pangolin played a central role in the pandemic
and how it was caused.

A study in India looked at the trade in pangolins and reported that
during a lockdown period (March to August 2020) there was a significant
increase in seizures compared to the same periods for the previous two
years. It seemed that pangolins were being targeted, or the lockdown
period was seen as an opportunity. The authors of this study concluded

Figure 6.2 An image that clearly associates the pangolin with the COVID-19 pandemic. Image from Shutterstock courtesy of Tolmindev, but has been made black and white.

that the increase in hunting and trade of these animals may have had an impact on their populations,[15] which is significant when you factor in that such populations were already threatened before COVID-19 emerged. Considering the effects on pangolins, it is ironic that others are saying that pangolins were not even involved in the COVID-19 outbreak.[16] However, such reports will not save the pangolin, whose reputation is probably tarnished, at least across much of the human population.

On the other hand, for the pangolin, it has been suggested that, in the longer term, the pandemic may have been a good thing. The pangolin has been brought to the attention of many who would not have considered the animal before, and the animal trade, having been blamed for the pandemic, may now be more closely monitored or even reduced.[17,18] China has banned the sale of pangolin scales for medical purposes.[19] Perhaps the pangolin will be safer after all, although they are still being traded and used in some countries, such as China.[20,21,22]

Other animals were also thought to be intermediaries between bats and humans.[23] Jie Zhao and his co-authors suggest that, as well as pangolins, there are several other species that could have passed the virus from bats to humans. It is worth noting that bats are still the source here, but intermediate species are grouped as either wild or domestic. The wild animals included, as well as the pangolin, are minks (although it is acknowledged here that this is a heavily farmed animal), turtles, snakes, and ferrets. Under domestic animals, they list yaks, dogs, cats, and pigs.

It is interesting here that reptiles are listed, as they were not found to be particularly susceptible in the bioinformatic analysis discussed above. For the turtle, there seems to be some contrary data to support or refute this idea,[24,25] and there seems to be a similar situation in the literature about the snake.[26,27] The focus seems to be turning to the raccoon dog.[28]

However, outside of the negativity surrounding bats and pangolins, there seems to be little evidence of any other animal species being targeted because of its involvement in the spread of the SARS-CoV-2 virus. This is, obviously, a good thing, but it may not be sustained in the future or during any future virus-causing epidemic. Any species that is deemed to be "at fault" for spreading a disease to humans may be vulnerable. Hopefully, international regulations may help. The Convention on International Trade in Endangered Species of Wild Fauna and Flora (CITES,[29] sometimes referred to as the Washington Convention) oversees the trade in animals and plants and ensures that endangered species are protected. The Nagoya Protocol[30] is concerned with the protection of genetic material, and this agreement regulates the movement of animal and plant materials across international borders. Such international regulations will be beneficial for animal welfare, especially as they are adopted by more and more countries. Animal trade and movements – including body parts – will be restricted, and hopefully the black market in such materials will also be reduced by vigilance by security personnel, including border control. Although it is important to note that these have been in place for some time, and poaching continues. Stopping poaching is a very complex task, including significant elements of behaviour change (including from purchasers), economic factors (such as poverty), environmental change (including land use pressure), and therefore legislation and regulation can only ever be a framework within which human action takes place.

6.3 Euthanasia/culling programmes that took place around the world

It has transpired that early in the pandemic the UK Government was considering whether the virus was being spread by pets, and they suggested that all cats in the United Kingdom should be euthanised,[31] all eleven million of them.[32] Fortunately, this was never enacted. There seemed to be no mention of dogs, so there must have been a perception that cats were of more concern.

As it emerged that COVID-19 was becoming a major issue for human health, one of the animal populations that was found to be infected was mink, and the focus centred on mink farming. Several countries have a significant trade in minks for their fur. In the European Union (EU), the countries most involved include Denmark, the Netherlands, Poland, and Spain. However, the breeding of minks for fur is a worldwide business, with farms in China and the United States. And it is big business. According to data from the *Fur for Animals* website,[33] in 2014, 42 million minks were killed in the EU. In Denmark this number was 18 million with 7.8 million in Poland. In China, a staggering 35 million animals were used in 2014 alone. Around the world, it has been estimated that in the 2013/2014 season over 87 million minks were used for pelts. This was estimated to be valued at over 3.7 billion euros. These minks are often held in tight, confined conditions, which is very poor for animal welfare, as shown in Figure 3.1. Furthermore, if the animals in such conditions become diseased, the infection can rapidly spread as the animals are in such close contact with each other. Such infections include, of course, COVID-19. If a human handler has COVID-19 and is also in contact with mink populations, such as in a farm, they may inadvertently – whether wearing PPE or not – infect an animal, and then the SARS-CoV-2 virus can jump from animal to animal, very efficiently it turns out. As already mentioned, there is some evidence that the virus could then "jump" back to the human population. And the final worry here is that the virus may have mutated during its travel through the animal population and then reinfect humans with a new variant. Hence, there were serious worries about the maintenance of the mink population.

Mink were doomed (in terms of immediate culling, rather than later being killed for their pelts) once the science showed that they were susceptible and could pass the virus back to us, so animal populations were culled. In Denmark, the government ordered the culling of 15 million mink across 1,139 farms (the number of farms highlights the scale of the industry). Prime Minister Mette Frederiksen said that this was needed to contain the SARS-CoV-2 virus.[34] In July 2020, the *Mirror* newspaper reported that 92,000 mink were to be culled at a farm in La Puebla de Valverde in Spain.[35] Over a thousand more mink were culled at another farm in Spain,[36] with a third incident seeing 3,100 animals gassed.[37] The science behind the mink issues in a farm in Spain was also reported in the science literature.[38]

Other countries were affected too. In the United States, 15,000 minks were said to have died, as reported on 10 November 2020.[39] In June 2020, the *Guardian* was reporting that in the Netherlands 10,000 minks were to be culled,[40] but shortly after, in July 2020, the same newspaper was reporting that over 1 million minks had been culled on Dutch farms.[41] The same article said that 87% of the minks had tested positive for SARS-CoV-2 and seven farm workers were also positive.

It can be seen that the scale of the problem in mink farms across the world was huge, and the outcome resulted in the death of millions of animals. It could be argued, of course, that all these animals were being bred for their pelts and would have been killed anyway, long before they got to old age. Therefore, was it such a bad thing? It meant that the pelts were not used, so the deaths were meaningless, but as pelts are only for fashion anyway, is there a big difference here? However, the long-term ramifications are more important perhaps. According to one source, up to 6,000 people lost their jobs due to the closure of mink farms in Denmark alone,[42] with a significant effect on many local economies. Many mink farms will not reopen. In Denmark, according to the *Guardian*, over 1,000 farms were forced to close, and many may never start again.[43] However, this might not be true. In December 2022, there was outrage when it was reported that Denmark was planning to import 10,000 minks to restart its pelt industry. These animals were thought to be sourced from several countries, including Norway, Spain, Poland, Finland, and even Iceland.[44] Therefore, as long as people wish to buy these furs and support the industry, there will be farms willing to supply that demand and the animals will continue to suffer. The effects of the pandemic may be seen as a blip in the industry when one looks back in a few years' time. It is a pity that this may not be a turning point, which would have reduced this unnecessary, and low-welfare, use of animals.

In the United Kingdom there is the Fur Farming (Prohibition) Act 2000,[45] which states that: "A person is guilty of an offence if he keeps animals solely or primarily—for slaughter (whether by himself or another) for the value of their fur [slightly paraphrased]." On a more positive note, for animal welfare, recently California has banned the sale of fur, making it the first state in the United States to make producing and selling fur illegal.[46] Therefore, in some parts of the world, there is an appetite for stopping the fur trade.

However, for the minks the story does not quite end there. There are political ramifications of what happened with the culling of minks. Back in Denmark, there was quite a lot of criticism of the decisions made regarding the mink trade during the pandemic, with a call for ministers to answer questions. This climaxed with the call for a general election, with the mink farming issue being reported as a major discussion point.[47]

Other animals, as well as minks, were affected by culling. There was a report in April 2020 from the United Kingdom of a four-month-old cat, called Ragdoll, that was put to sleep as it had become infected with SARS-CoV-2.[48] However, this sort of event seemed to be very rare. Of more significance were the reports of SARS-CoV-2 infections of hamsters in Hong Kong. Here, hamsters were thought to be reinfecting humans in pet stores, and, as a result, 2,000 hamsters were culled across the district.[49] Unlike the mink, these animals were destined to be pets and therefore may have lived a long and happy life, until COVID-19 got in the way. As part of the Chinese zero-COVID policy during 2022, there were calls in one Chinese district to cull all pets that belonged to people who were COVID-19 positive. Fortunately, after an outcry, the policy was scraped.[50] In Vietnam, 12 dogs owned by one couple were euthanised over fears that they would spread the virus. The owner was reported to have said, "My wife and I cried so much that we couldn't sleep."[51] It is not just the death of the animals that is a tragedy but the effect on the humans who care for them, which is also significant.

Culling has been proposed and, indeed, used as a method to control the spread of the SARS-CoV-2 virus. This is partly to stop the spread, particularly of variants, but also to try to prevent new variants springing up in animal populations and then getting back into the human communities. Whether this policy has been useful is a moot point, but millions of animals have been culled during the pandemic. Hopefully, this has now stopped. However, lessons do need to be learnt here. Either our breeding programmes for animals need an overhaul or we need to have better disease management. A simple knee-jerk reaction to kill any animal infected is not a sustainable situation. Firstly, we kill the innocent animals, which may not, in any case, be a significant risk to humans (for example, because they are not shedding significant viral load), but we are in danger of destroying peoples' companions (as in pets) or livelihoods (pet shops, etc.). And then restarting industries such as mink farming is also controversial, not least because such animals have to be transported internationally and often overseas where animal welfare in rough and cramped conditions should be questioned. Much of this industry is not needed (from a purely animal-centric and pragmatic view), and it is time we took a good look at whether the pandemic has given us an opportunity to take a step back and look hard at what we are doing with some animal species. Perhaps the United Kingdom, California, and other places need to show good practice here.

It was not just to control the virus that culling was suggested. In the United States there was a supply-chain issue for food supplies. Slaughterhouse availability and transport issues meant that animals were being culled at the farms, and therefore not used for food. At the end of April 2020, the *Guardian* newspaper was already reporting that over two million animals had been culled,[52] and that the president, Donald Trump, had intervened with an executive order to force slaughterhouses to stay open. As with the mink, these animals were being farmed for food, and, therefore, they would have died anyway, but being culled for no use was a sad thing to see. The talk was to "de-populate" farms, and this included pigs, cows, and chickens. The farms cannot economically keep the animals beyond the age where they are useful for human consumption, and, therefore, growing to an old age was not an option for these animals.

6.4 Animal behaviour, moving into populated areas (analogies to Chernobyl exclusion zone)

In 1986 (26 April) the Chernobyl power station exploded, in particular in reactor 4. This was a nuclear facility in the Ukraine, near Pripyat[53] north of Kyiv (often spelt as Kiev). Two people were killed immediately in the explosion, but many more died later of radiation effects. It was a world-affecting disaster. Sheep in Wales, for example, were affected, and this was thousands of miles away.[54] One of our professors at the time of the explosion was travelling across Europe (he was Austrian) and when he arrived back in Bristol, the mud from his car was investigated for radiation contamination, and there was no major surprise that the radiation was higher than expected from a normal road trip. Huge areas across Europe were affected, little or much, from the fallout. No doubt a similar situation arose in Japan when the Fukushima nuclear power station suffered a core meltdown in March 2011[55] (Figure 6.3).

Why are we discussing this here? In the aftermath of the Chernobyl explosion, 350,000 people were evacuated, and an exclusion zone was set up. This was eventually extended to cover 4,300 square kilometres. People were excluded, but nothing else. Animals and plants were allowed to remain in place, and, in some ways, this became a "paradise" for wild animals. With no humans to interfere with their lives, were the animals better off? This has been debated to some extent ever since the exclusion zone was set up.

There were several reports of the exclusion zone in Ukraine being a thriving ecosystem. Headlines suggest that the disaster "May Have Also Built a Paradise,"[56] with species such as bison and brown bears doing well. There have been impacts on the fish populations too. Lack of humans means fewer population in the water and less fishing. Some fish, such as catfish, have been found that are surprisingly large, but they have been able to live longer and eat well, although other fish species have been adversely affected.[57] However, others have disputed that the ecosystem is better, saying that there is little evidence of many species taking advantage of the exclusion of humans or that they are suffering from the effects of the radiation.[58] The situation is probably very species-dependent, with some animals being able to take advantage of the lack of human interference, but others suffering as a consequence. Ecosystems are complex, after all, and, as one animal thrives, another may be adversely affected.

Figure 6.3 A photograph of abandoned houses near the Chernobyl nuclear power station that exploded in April 1986. Photograph from Shutterstock, courtesy of Special View. Image has been made black and white.

It appears that some animals, especially if they are mobile, are able to take advantage of the absence of human activity. This is certainly what has been reported during the COVID-19 pandemic. *NPR* reported that "Humans travelled less during COVID restrictions. Animals travelled more."[59]

In many countries around the world, as already mentioned, there were lockdowns, where humans were forced by law to stay home and only leave for vital reasons, either for work or for getting food. This left regions feeling abandoned, and animals noticed. As can be seen in Figure 6.4, in Mitzpe Ramon in Israel, Nubian ibexes were seen roaming in the streets, a sight not normally seen. This was shown as a photo gallery in a UK newspaper.[60]

The Ibex in Israel were not an isolated case. *Country Living* had an article showing 13 examples of animals in strange places during lockdowns.[61] These included seabirds in Venice (Italy), Sika deer in Nara (Japan), rabbits in Christchurch (New Zealand), mountain goats in Llandudno (Wales, UK, as mentioned in the previous chapter), peacocks in Ronda (Spain) and in Dubai, deer in London (UK), monkeys and cows in New Delhi (India), geese in Adana (Turkey), pumas in Santiago (Chile), sheep in Samsun (Turkey), and ducks in Paris (France). Some of these cases do not sound that unusual, but, it may be reported because of increased numbers of animals seen, or their presence in particular locations where people would have previously been dominant. Certainly, the picture of

Figure 6.4 Nubian ibexes in the street in Mitzpe Ramon in the Negev desert (Israel) on 4 February during a national third lockdown. Photograph from Shutterstock courtesy of Gil Cohen Magen. Image has been made black and white.

the cow in India showing it lying in the middle of a highway flyover near Timarpur, with not a vehicle in the picture, does highlight the point of unusual animal spottings.

Other media outlets were reporting similar incidents. *RFI*[62] said that there were civets roaming around Kozhikode (State of Kerala, India), whilst the wild boar around Barcelona (Catalonia) were coming into the city. They, too, reported on the ducks in Paris. They also reported that dolphins were seen in several ports around the Mediterranean, which is unusual. Foxes were also seen in cities, although in some cities such as Bristol (UK) this is a common occurrence whether there is a pandemic or not.[63] However, the *RFI* article also pointed out that many birds are becoming more brazen and are seen more where humans would normally have scared them away, including sparrows and pigeons. Interestingly, they point out that because of the lack of boat activity the range of puffins has been seen to have expanded too. The BBC reported that kangaroos were seen in Adelaide (Australia) and deer were wandering into towns in Japan looking for food. They also reported squirrels "sunbathing" in California. Alligators were seen on beaches in South Carolina (United States), and coyotes were seen in San Francisco and even on the Golden Gate Bridge.[64] The BBC in a separate article[65] also reported that dolphins were seen in the Bosphorus (Istanbul, Turkey), where normally the high level of shipping would have kept them away, wild boar were seen in Haifa (Israel), while in Albania pink flamingo numbers were said to be increasing significantly. Others reported similar events with wild boar being seen in Bergamo (Italy) and in Paris (France).[66] Pink dolphins were returning to Hong Kong harbour and pelicans were sighted near Buckingham Palace in London.[67] It certainly seems as though in many places around the world animals have been taking advantage of the fact that humans were not around so much, and it was a shock to many how quickly this occurred once humans were confined indoors.

Of course, many of these animal behaviours will be short-lived. It may seem idyllic for an animal if a road or park is lacking in human activity, but as soon as a lockdown ends and the people swarm back into the streets, the wild animals will be frightened away and not return, unless another lockdown happens. However, some of the unusual animal behaviours may have longer-term effects, at least in the near future. The BBC article reported that there was a higher-than-expected number of baby leatherback sea turtles hatching in Phang Nga Province of Thailand. Assuming that a significant number of these babies will reach maturity, this would be good for the turtle population.

There have been some adverse effects of humans not being around. Many animals, including birds and mammals, rely on waste left behind by human activity as a food source.[68] During lockdowns, people were prevented from going to parks and recreational areas, and many birds and scavenging animals were deprived of their easily found meals. For example, stray animals in India have been reported to have suffered as there was no food or water for them, especially in the summer heat.[69] Many animal species would have suffered because the deliberate or incidental depositing of food was severely reduced.

Human activity has an inexorable influence on the behaviour of animals, and when the movement and actions of people are significantly altered, there is the inevitable knock-on effect with animal populations. Sometimes there is a benefit to animals, allowing them to roam more freely or breed more easily without interference, but often there is a detrimental effect. This balance was thrown akilter by the pandemic, and many strange or unexpected animal behaviours were reported. However, now that the pandemic is phasing out, it is unlikely that these events will persist, and animals will no doubt resort to the activities that were customary before the virus struck.

6.5 PPE being abandoned and the issues it causes

When the pandemic was known to be an issue for human health, there was a scramble around the world to source personal protective equipment (PPE) for human use. This was obviously a priority for hospitals, where patients with COVID-19, and therefore likely to spread the virus, were in high and dense numbers. Obviously, no one wished to be exposed, unnecessarily, to a virus that could lead to death. Also, even if there were no serious health consequences, hospital and other medical staff were forced to be off work so that they could recover and also so they would not spread the virus. Therefore, for such staff, there was an urgent need to wear PPE, and this included mouth masks, face masks, gowns, and gloves. Of course, all this equipment was for one-person use and for one-time use too. An enormous amount of equipment was needed and then thrown away.

Estimates of the UK Government's spending on PPE during the pandemic vary widely, and accurate figures are hard to come by. However, a report by the National Audit Office in November 2020 suggested that £15 billion had been budgeted for PPE for 2020–21,[70] and the final figure is likely far higher as this was by no means the end of the pandemic. By December 2020, the United States had spent an estimated seven billion US dollars on medical equipment, including PPE, and was passing legislation to spend ten billion more.[71] This pattern was repeated across the world, and of course all these staggeringly high numbers were official government purchases. On top of this was the huge spending by individuals, also desperate or mandated, to protect themselves when in public. For example, it was compulsory to wear a mask over one's nose and mouth in public spaces in the United Kingdom for months, and this included in shops and on public transport. There were exemptions to this, if one had a medical reason not to wear a mask, but this was not common and most people had to wear a mask. Many other countries around the world had more stringent laws on PPE, which persisted for considerably longer than the regulations in the United Kingdom.

The WHO also gave staggering numbers when reporting about PPE. They said that between March 2020 and November 2021, one and a half billion units of PPE were procured. This was equivalent to approximately 87,000 tonnes. To put that in perspective, it was said to be equivalent to "261 (747) jumbo jet aeroplanes" [brackets added for clarity]. It was stated that pyrolysis (subjecting the material to destructive heat) was the safest and best way to treat used PPE.[72] However, much of the PPE was discarded without such treatment and with little care to what happened to it.

For animals, face masks (nose and mouth cover) were the biggest issue, along with gloves. PPE from a hospital or industrial company would have been disposed of through official routes. In the United Kingdom, as elsewhere, this would have included being autoclaved, burnt, or similarly destroyed. Gloves would have been made primarily from vinyl or nitrile. Masks were more variable, with many people making their own. Commercial masks contained a variety of materials, whilst some homemade ones were simply single layers of cloth, and their efficacy was often questioned. When commercial masks were analysed, they were found to release heavy metals (lead, antimony, and cadmium) and plastic fibres into the environment.[73] Both have a major environmental impact. Heavy metals are toxic to animals[74] and plants,[75] both showing stress responses[76] when exposed to significant levels. In a similar vein, microplastics are also known to be harmful.[77] Microplastics are of particular relevance to aquatic systems,[78] food sources, and are a growing concern.[79] Of course, PPE is not the only source of microplastics, and it is said that they even come from face cleansers,[80] for example. However, although there is a wide range of sources of both heavy metals and microplastics that can contaminate the environment, the sheer scale of the use of PPE during the pandemic, and into the future, will leave toxic effects in a range of ecosystems, many of which we rely on for food, such as fish and other seafood.

There is, therefore, an unseen effect of the use of a large quantity of PPE. The manufacture, transport, and disposal will all add to the damage that humans are causing to the environment. However, there is also a visible effect of the PPE, which was used to combat the spread of the SARS-CoV-2 virus and, of course, other pathogens.

When walking down the street, how many times did you see PPE lying on the pavement or in the gutter? Certainly, it was a common sight in Bristol while commuting to work. No one was willing, not surprisingly, to pick this waste up and dispose of it in a bin. It was likely to be contaminated, and if it was not, it was not very nice or indeed sensible to handle. Therefore, it was left in place. It was going to degrade further, releasing its toxic waste into the environment.

WWF considered the potential scale of the problem and stated that:

> If even just 1% of the masks were disposed of incorrectly and perhaps dispersed in nature, this would result in as many as 10 million masks per month dispersed in the environment.[81] [auto-translation of site originally in Italian].

They went on to say that this would be equivalent to 40 thousand kilogrammes of plastic that would be deposited into the environment and then suggested that this was a situation that needed to be altered. Gabriel E. De-la-Torre et al.[82] looked at the littering by PPE along the coast in Lima, Peru, and estimated that the scale of the problem was up to 7.44×10^{-4} PPE m^{-2}, which does not sound like a lot until you think about how many square metres a beach may be.

How long does it take for PPE to degrade if abandoned in the environment? This is hard to give a precise answer to as it depends on the materials used in manufacture. However, some people have tried to give estimates. The *Ecoexperts* site[83] said that materials fabric may degrade quickly, but some of the filters in masks are made from polypropylene, which can take five hundred years to degrade. If you decided that a pair of leather gloves would give you protection from the virus, and then you were to lose them, it could take 50 years for the leather to degrade. For vinyl gloves the site states that it would "certainly take longer than an average human lifetime to see them disappear" and for nitrile they suggest it could take centuries. Therefore, any PPE left lying about in the environment is going to take a very long time to degrade and disappear.

Researchers have, not surprisingly, looked at the degradation of PPE materials. De-la-Torre et al., using simulated environmental conditions, found that sunlight helped to degrade materials and that chemicals were released into the water, as expected.[84] Similarly, Pizarro-Ortega et al.[85] found that microplastics were indeed released on PPE degradation. It takes a long time for these degradation processes to be completed.

If PPE does not disappear in the environment, what happens to it? It stays lying, or floating, about until it disappears. As can be seen in Figures 6.5 and 6.6, there is a likelihood that animals will be in the proximity of discarded PPE. Much of the PPE has straps, gaps, and hollows (such as finger holes), and such features are ideal for animal entrapment, as highlighted in Figure 6.7, where a bird has a face mask around its neck. Images of such results of discarded PPE may be rare, but the scale of the

Figure 6.5 A photograph of PPE sinking to the seafloor, in this case a tropical reef, with a turtle near that could easily become entangled. Photograph sourced from Shutterstock courtesy of Drew McArthur, but made black and white.

Figure 6.6 A fish avoiding, or about to get entangled, in PPE discarded during the COVID-19 pandemic. Photograph from Shutterstock, courtesy of Damsea, but has been made black and white.

Figure 6.7 A blackbird entangled in a disposable mask, discarded during the pandemic. Photograph from Shutterstock courtesy of JDzacovsky, but made black and white.

problem will be much larger than seen by people with cameras (even though many of us now have cameras on our phones).

Yang et al.[86] say that there was a danger from PPE from "entrapment, entanglement, and ingestion." They also discuss the impact on aquatic systems, the soil, atmosphere, and then on humans. They end by suggesting some recommendations, which we will revisit in Chapter 7.

On the other hand, Hiemstra et al.[87] found that some animals may be using PPE positively. They found that a bird in the Netherlands was using PPE as a nesting material, although they also reported a fish entangled in a glove in the canal. Interestingly, they have been collating a database of peoples' observations, which can be found at https://www.covidlitter .com/, which they describe as "The online -updated- table." Eighty-six events were listed and some of these data are shown in Table 6.1.

As well as the multiple events listed in Table 6.1, the database also lists many other examples. Some are repeats of the same species in the same country, located and reported by different people. But there are several other interesting examples: a common octopus hiding in a face mask in France; dogs playing with face masks; a macaque chewing a face mask in Malaysia; a kite (bird) carrying a face mask in the Netherlands. Many of these latter examples may seem quite innocuous, but it takes very little to go from chewing to ingesting, or from carrying to entanglement, and the negative ramifications of either of those events occurring (such as the serious intestinal blockages suffered by cows in India when they ingest plastic rubbish). What is also clear from these data is that face masks are by far the most likely cause of an issue, with few examples of gloves being a problem (although single events can be devastating for the animal involved). Numerous animal species are also affected here, including mammals (land and marine), birds, fish, and reptiles.

It is interesting that there are so many examples of animals using PPE for nesting material, and this is not just by birds, with rodents also getting in on the act. This is opportunism. Recently, it has been noted that birds (especially corvids) even use anti-bird spikes for nesting, suggesting that the birds may use it for nest protection![88] However, using human-derived materials such as PPE could also be dangerous if the young then become entangled or try to ingest the unnatural, and possibly toxic, materials. It is not unlikely that some of the discarded PPE may be contaminated with the SARS-CoV-2 virus. This may be of little issue to birds, but it may be a problem for rodents and bats.

What is clear is that this is a world-wide issue, with reports from the United States, Canada, Europe, Malaysia, Japan, and Australia, for example. Huge swathes of the world are not represented, however. This includes South America, Russia, and Africa, although there was a seal reported entangled in a face mask in Namibia. Therefore, the data are far from complete, and the data here rely on people, first, seeing an issue and, second, reporting it to the database. This is a great, rather worrying resource, but it is a mere snapshot of the real situation, and there seems little doubt that far more animals have suffered from discarded PPE than we will ever know.

Although the paper by De-la-Torre et al.[89] (2021) is about Lima in Peru, they have an interesting table that summarises what others have reported around the world. This includes finding PPE on beaches in Chile, Colombia, Argentina, and Kenya. They also list PPE in rivers in

Table 6.1 Some of the data extracted from the "The online-updated-table" regarding PPE effects on animals

Effect seen	Type of PPE	Animal affected	Country reported
Entanglement	Face mask	American robin (*Turdus migratorius*)	Canada
	Face mask	White tailed eagle (*Haliaieetus albicilla*)	Germany
	Face mask	Mute swan (*Cygnus olor*)	UK
	Face mask	Gull sp. (*Laridae sp.*)	UK
	Face mask	Peregrine falcon (*Falco peregrinus*)	UK
	Face mask	Checkered pufferfish (*Sphoerides testudineus*)	USA
	Face mask	Red fox (*Vulpes vulpes*)	UK
	Face mask	Mallard (*Anas platyrhychos*)	Italy
	Face mask	Shore crab (*Carcinus maenas*)	France
	Face mask	Serotine bat (*Eptesicus serotinus*)	The Netherlands
	Face mask	European hedgehog (*Erinaceus europaeus*)	The Netherlands
	Face mask	Mallard (*Anas platyrhychos*)	UK
	Face mask	Gull sp. (*Laridae sp.*)	The Netherlands
	Face mask	Common merganser (duck) (*Mergus merganser*)	UK
	Face mask	Herring gull (*Larus argentatus*)	Canada
	Face mask	Coral sp.	Philippines
	Face mask	Mallard (*Anas platyrhynchos*)	The Netherlands
	Face mask	Mute swan (*Cygnus olor*)	Ireland
	Face mask	Puffin (*Fraticula arctica*)	Ireland
	Face mask	Brown treecreeper (*Climacteris picumnus*)	Australia
	Glove	European hedgehog (*Erinaceus europaeus*)	UK
Entrapped	Glove	Perch (*Perca fluviatilis*)	The Netherlands

(Continued)

Table 6.1 (Continued) Some of the data extracted from the "The online -updated- table" regarding PPE effects on animals

Effect seen	Type of PPE	Animal affected	Country reported
Ingested	Face mask	Dog (*Canis lupus familiaris*)	UK
	Face mask	Dog (*Canis lupus familiaris*)	USA
	Face mask	Dog (*Canis lupus familiaris*)	The Netherlands
	Face mask	Herring gull (*Larus argentatus*)	UK
	Face mask	Green sea turtle (*Chelonia mydas*)	Japan
	Face mask	Green sea turtle (*Chelonia mydas*)	Australia
	Face mask	Common dolphin (*Delphinus delpis*)	Tenerife
	Face mask	Yellow-legged gull (*Larus michahellis*)	Italy
	Face mask	Crow (*Corvus corone*)	The Netherlands
Nesting material	Face mask	Common coot (*Fulica atra*)	The Netherlands
	Face mask	House finch (*Haemorhous mexicanus*)	USA
	Face mask	House mouse (*Mus musculus*)	Germany
	Face mask	Crow (*Corvus corone*)	The Netherlands
	Face mask	Squirrel (*Sciuridae sp.*)	India
	Face mask	White stork (*Ciconia ciconia*)	Germany
	Face mask	Little penguin (*Eudyptula minor*)	Australia

Data from https://www.covidlitter.com/ (accessed 09/01/23).

Colombia and Indonesia and in a lake in Ethiopia. Urban or city discarded PPE were seen in numerous cities, with the countries involved listed as Brazil, Kenya, Canada, Portugal, and Nigeria. This is no surprise as it is likely that PPE could be found in most countries around the world if one were to look hard enough, but it does show that several research groups are monitoring the issue, as eight citations (including their own) are given.

Therefore, from a skim through the literature and data, it is clear that the amount of PPE discarded into the environment can have a significant impact on many animal species, and any future pandemic management needs to be cognisant of this indirect effect of the vast amount of PPE produced and then used as people travel around their communities.

However, as with all the discussions about the impact of the pandemic, such issues as PPE being discarded have to be put into perspective. Yes, there have been negative effects of PPE use, with animals suffering and dying, but this is on top of the ongoing issue of waste dumped in the environment. It has been estimated that 75% of plastic waste in the environment is composed of only ten different types of products, with the top four being single-use bags, plastic bottles, food containers, and food wrappers. This accounts for approximately 50% of the plastic waste found in the sea.[90]

According to a website called *Litter*,[91] "9 billion tons of litter ends up in the ocean every year." They go on to say that 50% of the litter is cigarette butts, but the most commonly found items during cleaning are waste from takeaway food outlets. All this waste will be a hazard to animals that may try to eat it or get entangled in it. It is on top of this that pandemic-generated PPE waste is deposited.

To combat the waste in the environment which is generated during a pandemic, a cultural change about discarding of litter of all kinds is needed. Reports suggest that men are more likely to drop litter than women, and that the age group most likely to be involved is 18–34. Targeting such demographics may be needed to reduce the amount of litter left in the environment, including that used during a pandemic. Furthermore, provision for people to deposit waste, which is then dealt with properly, needs to be put in place by governments, both on national and local scales.

6.6 Massive use of cleaning/disinfectant materials, etc. and impact on the environment

One of the ways used to combat the transmission of the SARS-CoV-2 virus was the widespread use of disinfectants. It was thought that the virus was spread by contact, that is, it could be picked up from surfaces. In March 2020 the WHO said that: "COVID-19 virus is primarily transmitted between people through respiratory droplets and contact routes."[92] However, others have disputed this. Dyani Lewis, writing for *Nature*, says that the virus "can linger on doorknobs and other surfaces, but these aren't a major source of infection."[93] There is some evidence that the virus can linger on plastic and stainless-steel surfaces,[94] but this will, of course, depend on how long they are there as to whether they remain infectious. The SARS-CoV-2 virus is an envelope virus and can easily and relatively quickly dry out and disintegrate. According to *Medical News Today*,[95] quoting from the *New England Journal of Medicine*,[96] the virus could persist for 72 hours on the surface of steel and plastic, 24 hours on cardboard, but only four hours on a copper surface. The same article also discussed how long the virus could last in the air, suggesting that as an aerosol the virus may last longer than three hours in air, but this is dependent on the temperature and humidity. Others have also published data on this. Corpet, writing in the journal *Medical Hypotheses*, stated that the virus "is inactivated much faster on paper (3 hours) than on plastic (7d)" and then went on to say that water droplets on surfaces may aid in the virus persisting.[97] Others say the preservation of the virus depends on the variant, with Omicron being particularly persistent.[98]

All these reports suggest that SARS-CoV-2 can persist in the environment, and this made people worried. It was quickly known that the virus would not resist detergents and other cleaning fluids, such as disinfectant-containing hand washes, and therefore it seemed like the sensible thing to do was to treat any potentially infected surfaces with detergents, with what some people called a deep-cleaning regime. Hands were supposed to be washed for at least 20 seconds,[99] with a thorough covering of soap. Lots of people, of course, ignored such advice. Cleaning in hospitals, as well as in commercial and domestic settings, was encouraged. However, in some places this was taken to extremes, and the environment was cleaned too, as can be seen in Figures 6.8 and 6.9. In the Philippines (Manila on 11 March 2020), antiseptic fluids were sprayed onto the streets from fire

Figure 6.8 A photograph of people spraying cleaning materials in a garden. Such use of disinfectants was commonly practised during the COVID-19 pandemic. Photograph sourced from Shutterstock courtesy of Art Photo, but made black and white.

Figure 6.9 Disinfectants being sprayed from a fire truck in Guwahati, Assam, India. Image from Shutterstock courtesy of Talukdar David, but made black and white.

trucks.[100] Similar tactics were used in the streets in Tehran, Iran. In the streets of Wuhan, China, workers were spraying disinfectants in the residential areas from hand-held devices, whilst people leaving a make-shift hospital in the same city were being sprayed outside before they were allowed to leave. The same report shows the Rialto Bridge in Venice being disinfected, along with a bus station in Gwangju, South Korea. Escalators can be seen being sprayed (in Tbilisi, Georgia) as well as the outside of a factory (Huzhou, Zhejiang Province, China). Many more examples, both indoors and outdoors, are shown in photographs. Some people have even discussed the use of drones for spraying large areas,[101,102] which may not have been necessary, although such technology could have positive advantages in a pandemic, such as delivering medical supplies, in surveillance, and in communications.[103]

Measures such as wide-scale spraying of disinfectant fluids may have been good to protect people from the virus, but at the end of the day all these chemicals had to go somewhere. They would have been washed into drains and water courses, and much of it would not have been gathered into a treatment works before entering streams and rivers (even in countries where there is a water treatment infrastructure, such as the United Kingdom, this is often not the case[104]). There would have been inevitable consequences for the ecosystems in the vicinity of spraying, and much farther away as the chemicals were carried down rivers and eventually to the sea.

In a recent paper about the use of disinfectants during the COVID-19 pandemic, Shakeel Ahmad Bhat et al.[105] state:

> people used it [chemicals] aggressively because of panic conditions, anxiety and unconsciousness, which can have a detrimental impact on human health and the environment. Our water bodies, soil and air have been polluted by disinfectants, forming secondary products that can be poisonous and mutagenic.

They go on to say that they should be used in such a way as to have "sufficient precautions to minimise exposure to harmful by-products."

However, as with all aspects of the discussion about the effects of COVID-19, the use of widespread chemicals needs to be put into perspective. For example, in the winter, in many countries, salt is applied to roads to stop them from icing over and becoming slippery. However, as well as having significant impacts on travel safety, the use of sodium chloride has a range of very negative effects. It can damage the vehicles themselves, especially affecting steering and braking efficiencies. The salt can damage the road surface, and it can cause corrosion to bridges. It can also cause a major environmental problem. The salt gets sprayed onto land adjacent to roads, and many plants are not salt tolerant. The salt also has to end up somewhere, and much of it runs off into streams and rivers, where freshwater animals and plants are affected.[106] Even in countries that have large national parks, such as Canada, the use of salt as a de-icing agent has not been stopped.[107]

To try to combat the use of salt, alternatives have been developed. One interesting de-icer now being used is made from plants, in particular beets, from which a juice is made and sprayed on roads and paths.[108] The article cites a press release from *Morton Arboretum,* which says:

> Beet juice is an effective alternative to salt alone because it lowers the freezing point of water to as low as -20 degrees [C].[109]

However, as with all clever new ideas there are caveats, and there have been concerns that the use of this product may not be good for insects, as it "causes fluid retention and alters organ function in mayflies."[110]

The pandemic saw widespread use of disinfectants, and these undoubtedly were not good for the environment, but unless more eco-friendly alternatives are developed, or they are used less, future pandemics will no doubt see the same activities, with more harm to our ecosystems.

6.7 The effect on companion animals, including dogs, cats, and rodents

Did your neighbour suddenly adopt a dog? Many did in the United Kingdom. However, was this a real effect or did it just seem that way?

Jeffery Ho et al.[111] looked at the interest in pet adoption during the pandemic and they stated: "In conclusion, the global interest in pet adoptions surged in the early phase of the pandemic but [was] not sustainable." Interest in dogs was more transient than that for cats. However, their abstract ended with a warning:

> With the launch of COVID-19 vaccines, there is a concern for separation anxiety and possible abandonment of these newly adopted pets when the owners would leave their homes for work in the future.

According to *The Dog People*[112] pet adoptions were significantly increased during the pandemic, so much so that they say that the term "pandemic puppies" was used, and they say that "nearly 40% of British people welcomed a new dog during the COVID-19 pandemic." It was not just dogs, but cats were welcomed too, with 39% of adopters having a cat and 8% having both. Liat Morgan et al. tend to agree.[113] In their abstract they say, "the interest in dog adoption and the adoption rate increased significantly, while abandonment did not change." According to a report in the *Guardian* newspaper (January 2022)[114] in the USA, nearly nine million dogs were adopted since the start of the pandemic, which they equate to the population of New York City, meaning that for every three humans in the United States there is now a dog. Cats were also adopted in increased numbers.

So why did this happen? Many people, either in families or on their own, looked for companionship, and many thought that they now had time, during lockdowns, to look after a pet. No doubt there was pressure from children too.

Megan K. Mueller et al. looked at the relationship between the ownership of companion animals and loneliness in adolescents.[115] Dog owners had less loneliness before the pandemic. People reported that they spent more time with their pets and that this helped to cope with stress. In Malaysia, Dasha Grajfoner et al.[116] found that pet ownership, compared to not owning a pet, increased mental health. Pet owners were able to deal with adverse situations better and had more positive emotions, but there was little difference in how the pet owners and people without pets coped with stress or depression. They concluded that owning a pet during the pandemic was a good thing to do. In Singapore, Tan et al.[117]

seemed to agree, concluding the highlights of their paper by saying, "Pet ownership and attachment play important roles in enhancing physical and mental health during a lockdown period."

It seems, therefore, that owning a pet during the pandemic was a good thing to do.

However, not all studies agreed. In Australia there was a survey called *COvid-19 and you: MentaL HeaLth in Aus Tralia now survEy* or COLLATE. Looking at this data, A. Phillipou and colleagues state:

> Contrary to expectations, the findings suggest that during a specific situation such as a pandemic, pets may contribute to increased burden among owners and contribute to poorer quality of life.[118]

The jury was also out for others. In their paper, Ece Beren Barklam and Fatima Maria Felisberti said that the effect of pet ownership depended on the person involved. Those with low resilience might benefit, but those with naturally high resilience might be worse off.[119] Some have coined the term "The Pet Effect Paradox," as it was expected that pet ownership would, of course, have a positive effect during COVID-19, but many studies showed little or no effect.[120]

Many of these studies are being published as the pandemic wanes, as it takes time to gather, analyse, and publish the data (the peer review process can take weeks or even months in some cases). At the time, when the pandemic was in full flow, the wisdom was that pets were a good thing to have. There was little negative feeling, except for maintenance costs and the time needed to look after a pet properly. It would give companionship, particularly in a locked-down house, and it would be an excuse to go out and get some exercise (at least for dog owners). There were few reports of pets becoming infected or passing the virus on. They were safe, or so people thought. Therefore, many people decided that a dog or cat was for them. It appears that the increase in well-being expected was not universally seen.

So, what happened to these pets as the pandemic unfolded and then faded?

To start with, as the pandemic unfolded and people were told to go into lockdown, many people adopted animals, and many of these were from rescue shelters, and it was thought to be a good thing. Unwanted animals were suddenly in demand. But as time went on, there was a fear that pets would be abandoned, either because of the worry of animals spreading the virus or because pets were no longer wanted and were becoming a burden once people returned to normal life. There were reports of pets being abandoned in Wuhan during the pandemic,[121] for example. The same paper concludes that during the pandemic, pet abandonment is "neither justified nor morally supported."

There are obviously circumstances when pets are left without an owner. With the high levels of SARS-CoV-2 infection in the human population, there were obviously high numbers of patients, and many of these are hospitalised, and many died. Therefore, pets are left, needing either short-term care or permanently rehoming. In the United States, *CNN* ran an article entitled, "This is what happens to the pets left behind when their owners die from coronavirus," describing animals being rescued from apartments and then taken to animal shelters.

However, as normality returned, people mainly go back to the way of life that they had before the pandemic, and, in many cases, these were without a pet. In the United Kingdom, the Royal Society for the Prevention of Cruelty to Animals (RSPCA) reported that there was a 25% increase in the abandonment of pets. This may not have been completely because of the end of the pandemic, although that was likely to be a contributing factor. Shortly after many countries were thinking that the pandemic was over, there was the start of a war that had global consequences. On 24 February 2022, Russia invaded Ukraine.[122] This was devastating for those concerned, but it also made energy costs spike, and the cost of living for many people rose enormously, far more than anyone had predicted. The finances of many households were squeezed, and pets were a financial burden, with feeding costs and vet fees and possible insurance costs. It was not a surprise that many pets were deemed too much to keep, and this led to increased abandonment of animals.[123] The RSPCA even posted a "Meet the unwanted lockdown puppies"[124] as they were expecting a surge in abandonment and were hoping to pique peoples' interest in adoption of any animals brought in.

The BBC reported that 3.2 million pets were bought in the United Kingdom during lockdowns, but that abandonment was now significantly increased. One rescue centre said that the number of dogs being brought to them was the highest it had been in 15 years. It reported that people were pretending that some dogs were strays and had been found, just as an excuse to get rid of them. One had been advertised for sale, and then when it was not sold, was brought to the rescue centre as a stray.[125] The BBC even reported a phrase to describe such activity: "Dogfishing." This is defined as "misleading someone into buying a dog which may not be what it seems."[126]

National newspapers were reporting the high levels of pets' abandonment, with one saying the owners neither had the time nor the money to look after them anymore.[127] This was clearly a significant issue for animal welfare. Although many animals were sold or taken to rescue centres, many were simply kicked out of the home to fend for themselves. One of the oddest examples was the abandonment of pigs. Apparently, pigs' ownership as pets increased during the pandemic, but now these animals are being abandoned too. A rescue centre called *My Lovely Pig Rescue* reported a massive increase in pigs it needed to care for, saying that they had even had reports of pigs being thrown over a wall, while others were abandoned in bogs and forests, and many of these animals may have been pregnant as neutering pigs was not commonly carried out.[128]

There were also some problems with pet ownership during the pandemic. It was hard, or impossible, to travel to a veterinary surgeon during a lockdown. Vets themselves may have been off work suffering from COVID-19, or another illness, as doctors too were struggling to cope with demand. As well as the short-term effects on the health of the animals, there were also reports that there were likely to be longer-term effects. For example, many animals were not spayed and neutered during the pandemic.[129] In the United States during 2020 and 2021, veterinarians carried out 2.7 million fewer such surgeries, and it is feared that this will lead to more animal euthanasia in the future. The conclusion of a scientific paper about this suggested that spaying/neutering capacity needs to be increased to mitigate the issue.[130]

As people returned to work, pets were left in homes, which during the day were now empty. These animals suffered from loneliness and feelings of being abandoned, what has been termed as "separation anxiety." Animals that were spending 24 hours a day in the company of humans suddenly had to be on their own for extended periods of time. The mental health of such animals suffered, and it was suggested that owners prepare their pets for this eventuality by training and increasing the periods of being alone gradually before a full return to work was required.[131]

Anecdotally, one of the consequences of the pandemic was the poor animal training that could take place. Because of the prohibition of travel, as well as social distancing guidance, many of the dog training services were curtailed. This meant that many people who were new to pet ownership had little opportunity to access professional help and dogs were not well trained. Furthermore, such animals had had limited opportunity to mingle with other people and other dogs, as walking and running free were severely restricted. If these dogs then ended up in shelters and were eventually given to new owners, these new owners felt that the level of obedience of these animals was below the standard that they would expect if the first owner had trained and walked the dogs properly.

As the effects of the pandemic hit, many people clearly turned to pet ownership as a way to cope with stress, anxiety, and potentially loneliness because of long periods of lockdown. But it seems as though pet ownership was not a positive experience for many owners, and it certainly has led to numerous problems for animals. Many are now not wanted. No doubt people will not remember the negative sides of pet ownership when the next pandemic arrives, but perhaps better education around this issue will help to stop unwanted animals from suffering in the future.

It makes an interesting footnote to this section that recently there was published a point of view that if one cares about animal welfare at all, then one should not own a pet at all.[132] Troy Vettese says, "Dogs lead lives of loneliness. Grey parrots die years earlier than their natural lifespans. And it is hard to fathom the boredom of pet fish." Apparently, New York State has banned the sales of dogs, cats, and rabbits in pet stores in an attempt to increase animal welfare, stopping mass breeding for the pet market and encouraging people to adopt from shelters instead. If such a policy became more widespread, perhaps in the next pandemic, pets will be more difficult to obtain. This might not be a popular view, but it is an interesting take on how the pet industry may move in the future.

6.8 Conclusions

The pandemic had a range of indirect effects on a wide variety of animals. These are summarised in Table 6.2.

COVID-19 almost certainly originated in bats, but it soon became known as a disease of humans. As discussed previously (Chapter 3), many animals suffered as a consequence of becoming infected, but there were a range of indirect effects too, many of which were not well publicised as the pandemic unfolded. Animals in some places were persecuted as they were deemed to be sources of either SARS-CoV-2 or possible related viruses. Many people do not like bats anyway, and the pandemic simply made this perception worse. Several animals were mooted as intermediates between bats and humans, and they were not well tolerated either.

It was human action in trying to stay healthy and safe during the pandemic that had the largest effect. Many animals were culled, especially mink in farms, with numbers of euthanised animals running into the millions. Both PPE and disinfectants were used in vast quantities and much of this ended up in the environment, with the subsequent damage to ecosystems. Both animals and plants were harmed, as no doubt were the prokaryotes[133] that we cannot see. Some of these effects may be short-lived as ecosystems recover, but much of the PPE will persist in the environment for a very long time, as plastics and other materials slowly disintegrate.

Pets were seen as a good thing as the pandemic unfolded, and there were reports of rescue centres struggling to keep up with demand as people looked to adopt a dog or cat. However, it was not all good news for the pets. Many were later abandoned or suffered from isolation as people returned to work after lockdowns.

Table 6.2 Summary of some of the effects of SARS-CoV-2 on animals

Effect on animals	Direct/ indirect effect	Outcome	Examples of effects
Infection with symptoms	Direct	Recovery of health	Numerous animals infected, and some died, e.g., lions
Animals targeted	Indirect	Animals harmed	Bats and pangolins
Animals roaming more	Indirect	Animals extending their habits	Deer, puma, etc.
Culling/euthanasia	Indirect	Animals dying	Mainly minks, but also hamsters and cats
Discarding of PPE	Indirect	Animals entangled and PPE ingested	Many animals, birds, and marine life affected, with some deaths
Overuse of detergents/ disinfectants	Indirect	Damage to the environment	Rivers, streams, and marine environments
Abandonment	Indirect	Animals left to fend for themselves	Pets abandoned
Increase in domestication	Indirect	More adoption of pets	e.g., dogs and cats

However, it was not all bad news for animals. Many areas were temporarily abandoned by humans as lockdowns hit, and many animals took advantage of this. Unfortunately for them, it was not long before the humans returned.

There is much that can be learnt here about how human activity during a pandemic has indirect effects on a range of animals, from redesigning PPE to using fewer chemicals. This will be further discussed in the next chapter, when we look to the future, as further epidemics are only just over the horizon.

Notes

1 Moreira-Soto, A., Taylor-Castillo, L., Vargas-Vargas, N., Rodríguez-Herrera, B., Jiménez, C. and Corrales-Aguilar, E. (2015) Neotropical bats from Costa Rica harbour diverse coronaviruses. *Zoonoses and Public Health*, *62*, 501–505.
2 Moreira-Soto et al. (2015).

3 Lo, V.T., Yoon, S.W., Noh, J.Y., Kim, Y., Choi, Y.G., Jeong, D.G. and Kim, H.K. (2020) Long-term surveillance of bat coronaviruses in Korea: Diversity and distribution pattern. *Transboundary and Emerging Diseases*, *67*, 2839–2848.

4 Smith, I. and Wang, L.F. (2013) Bats and their virome: An important source of emerging viruses capable of infecting humans. *Current Opinion in Virology*, *3*, 84–91.

5 Wang, L.F., Shi, Z., Zhang, S., Field, H., Daszak, P. and Eaton, B.T. (2006) Review of bats and SARS. *Emerging Infectious Diseases*, *12*, 1834.

6 Voigt, C.C., Grasse, P., Rex, K., Hetz, S.K. and Speakman, J.R. (2008) Bat breath reveals metabolic substrate use in free-ranging vampires. *Journal of Comparative Physiology B*, *178*, 9–16.

7 Johnson, N., Aréchiga-Ceballos, N. and Aguilar-Setien, A. (2014) Vampire bat rabies: Ecology, epidemiology and control. *Viruses*, *6*, 1911–1928.

8 Quora: Do I need to be worried about bats flying around at dusk? https://www.quora.com/Do-I-need-to-be-worried-about-bats-flying-around-at-dusk-I-was-standing-outside-on-my-lawn-last-night-when-I-noticed-a-few-bats-flying-overhead-I-didn-t-feel-any-bats-fly-into-me-but-now-I-m-paranoid-that-I-could (Accessed 06/01/23).

9 Zhao, H. (2020) COVID-19 drives new threat to bats in China. *Science*, *367*, 1436–1436.

10 *Ecological Killing is Under Heated Debate—Revision of Wildlife Protection Law Must Involve Experts from all Related Fields* (2020): https://xw.qq.com/cmsid/20200214A0JB1X00 [in Chinese: cited in Zhao (2020)].

11 Penn Political Review: The pigs of Cairo: Swine Flu, garbage people, and the cull of 2009: https://pennpoliticalreview.org/2020/12/the-pigs-of-cairo-swine-flu-garbage-people-and-the-cull-of-2009/ (Accessed 05/06/23).

12 Koopmans, M., Daszak, P., Dedkov, V.G., Dwyer, D.E., Farag, E., Fischer, T.K., Hayman, D.T., Leendertz, F., Maeda, K., Nguyen-Viet, H. and Watson, J. (2021) Origins of SARS-CoV-2: Window is closing for key scientific studies. *Nature*, *596*, 482–485. https://doi.org/10.1038/d41586-021-02263-6.

13 Nature: Did pangolins spread the China coronavirus to people? https://www.nature.com/articles/d41586-020-00364-2 (Accessed 06/01/23).

14 Animal Parks: Save the pangolins: https://www.africanparks.org/save-pangolins?gclid=EAIaIQobChMIp5D9zYOz_AIVhO_tCh0zNQCoEAAYBC AAEgKvOfD_BwE (Accessed 06/01/22).

15 Aditya, V., Goswami, R., Mendis, A. and Roopa, R. (2021) Scale of the issue: Mapping the impact of the COVID-19 lockdown on pangolin trade across India. *Biological Conservation*, *257*, 109136.

16 Frutos, R., Serra-Cobo, J., Chen, T. and Devaux, C.A. (2020) COVID-19: Time to exonerate the pangolin from the transmission of SARS-CoV-2 to humans. *Infection, Genetics and Evolution*, *84*, 104493.

17 Guardian: Covid-19 – a blessing for pangolins? https://www.theguardian.com/environment/2020/apr/18/covid-19-a-blessing-for-pangolins (Accessed 06/01/23).

18 PBS: COVID-19: A catastrophe or opportunity for pangolin conservation?: https://www.pbs.org/wnet/nature/blog/covid-19-pangolin-conservation/ (Accessed 06/01/23).

19 Mongabay: Banned: No more pangolin scales in traditional medicine, China declares: https://news.mongabay.com/2020/06/banned-no-more-pangolin-scales-in-traditional-medicine-china-declares/ (Accessed 06/01/23).

20   Guardian: China still allowing use of pangolin scales in traditional medicine: https://www.theguardian.com/environment/2020/oct/13/china-still-allowing-use-of-pangolin-scales-in-traditional-medicine (Accessed 17/07/23) (posted 13/10/20).

21   Environmental investigation agency: Will China's refusal to close its domestic markets for pangolins mean the end of the species? https://eia-international.org/blog/will-chinas-refusal-to-close-its-domestic-markets-for-pangolins-mean-the-end-of-the-species/ (Accessed 17/07/23).

22   China Dialogue: Product labels may be failing to protect pangolins in China: https://chinadialogue.net/en/nature/product-labels-may-be-failing-to-protect-pangolins-in-china/ (Accessed 17/07/23).

23   Zhao, J., Cui, W. and Tian, B.P. (2020) The potential intermediate hosts for SARS-CoV-2. *Frontiers in Microbiology, 11,* 580137.

24   Liu, Z., Xiao, X., Wei, X., Li, J., Yang, J., Tan, H., et al. (2020) Composition and divergence of coronavirus spike proteins and host ACE2 receptors predict potential intermediate hosts of SARS-CoV-2. *Journal of Medical Virology, 92,* 595–601.

25   Luan, J., Jin, X., Lu, Y. and Zhang, L. (2020) SARS-CoV-2 spike protein favors ACE2 from Bovidae and Cricetidae. *Journal of Medical Virology, 92,* 1649–1656.

26   Ji, W., Wang, W., Zhao, X., Zai, J. and Li, X. (2020) Cross-species transmission of the newly identified coronavirus 2019-nCoV. *Journal of Medical Virology, 92,* 433–440.

27   Zhang, C., Zheng, W., Huang, X., Bell, E.W., Zhou, X. and Zhang, Y. (2020) Protein structure and sequence reanalysis of 2019-nCoV genome refutes snakes as its intermediate host and the unique similarity between its spike protein insertions and HIV-1. *Journal of Proteome Research, 19,* 1351–1360.

28   Nature: COVID-origins study links raccoon dogs to Wuhan market: What the scientists think: https://www.nature.com/articles/d41586-023-00827-2 (Accessed 17/07/23).

29   CITES: https://cites.org/eng (Accessed 06/01/23).

30   UK Government: Regulations: The Nagoya Protocol on access and benefit sharing (ABS): https://www.gov.uk/guidance/abs (Accessed 06/01/23).

31   Guardian: UK cat cull was considered early in Covid crisis, ex-minister says: https://www.theguardian.com/world/2023/mar/01/uk-cat-cull-was-considered-early-in-covid-crisis-ex-minister-says (Accessed 07/03/23).

32   MSN: Government considered cull of all 11,000,000 cats in the UK at start of Covid pandemic: https://www.msn.com/en-gb/news/world/government-considered-cull-of-all-11000000-cats-in-the-uk-at-start-of-covid-pandemic/ar-AA187HuI (Accessed 07/03/23).

33   Fur for Animals: Mink facts: https://respectforanimals.org/facts-about-animals-used-for-their-fur/mink-facts/ (Accessed 06/01/23).

34   Fox29 Philadelphia: Millions of minks ordered to be euthanized after mutated COVID-19 virus spread to humans: https://www.fox29.com/news/millions-of-minks-ordered-to-be-euthanized-after-mutated-covid-19-virus-spread-to-humans (Accessed 06/01/23).

35   Mirror: Spain orders deaths of 92,000 mink after workers and animals catch coronavirus: https://www.mirror.co.uk/news/world-news/spain-orders-deaths-92000-mink-22368290 (Accessed 06/01/23).

36   Murcia Today: 1100 mink culled in northern Spain after one tested positive for Covid: https://murciatoday.com/1100-mink-culled-in-northern-spain-after-one-tested-positive-for-covid_1554527-a.html (Accessed 06/01/23).

37 Murcia Today: Galicia mink farm closed after detection of another covid outbreak: https://murciatoday.com/archived-_-galicia-mink-farm-closed-after-detection -of-another-covid-outbreak_1609434-a.html (Accessed 06/01/23).

38 Badiola, J.J., Otero, A., Sevilla, E., Marín, B., García Martínez, M., Betancor, M., Sola, D., Pérez Lázaro, S., Lozada, J., Velez, C. and Chiner-Oms, Á. (2022) SARS-CoV-2 outbreak on a Spanish mink farm: Epidemiological, molecular, and pathological studies. *Frontiers in Veterinary Science*, 8, 805004.

39 Reuters: Coronavirus kills 15,000 U.S. mink, as Denmark prepares for nationwide cull: https://www.reuters.com/article/us-health-coronavirus-usa-minks-idUSK-BN27Q35V (Accessed 06/01/23).

40 Guardian: Dutch farms ordered to cull 10,000 mink over coronavirus risk: https://www.theguardian.com/world/2020/jun/06/dutch-mink-farms-ordered -to-cull-10000-animals-over-coronavirus-risk (Accessed 06/01/23).

41 Guardian: A million mink culled in Netherlands and Spain amid Covid-19 fur farming havoc: https://www.theguardian.com/world/2020/jul/17/spain-to-cull -nearly-100000-mink-in-coronavirus-outbreak (Accessed 06/01/23).

42 BBC: Fur industry faces uncertain future due to Covid: https://www.bbc.co.uk /news/business-55017666 (Accessed 05/06/23).

43 Guardian: Ghost farms: The mink sheds abandoned to the pandemic: https:// www.theguardian.com/environment/2022/nov/14/ghost-farms-the-mink -sheds-abandoned-to-the-pandemic (Accessed 06//01/23).

44 Humane Society International: Denmark's plan to restart mink fur farming by importing mink is branded contemptible by Humane Society International/ Europe: https://www.hsi.org/news-media/denmarks-plan-to-restart-mink-fur -farming-by-importing-mink-is-branded-contemptible-by-humane-society -international-europe/ (Accessed 06/01/23).

45 UK Government: Fur Farming (Prohibition) Act 2000: https://www.legislation .gov.uk/ukpga/2000/33/section/1 (Accessed 06/01/23).

46 Hunger: California officially bans fur sales – but mass spenders like China are still miles behind: https://www.hungertv.com/editorial/california-officially -bans-fur-sales-but-mass-spenders-like-china-are-still-miles-behind/ (Accessed 06/01/23).

47 Guardian: Danish general election called after PM faces mink cull ultimatum: https://www.theguardian.com/world/2022/oct/05/denmark-prime-minister -mette-frederiksen-general-election (Accessed 06/01/23).

48 Mail online: Eight lions test positive for Covid-19 at an Indian zoo in the first case of its kind in the country: https://www.dailymail.co.uk/news/article -9540377/Eight-LIONS-test-positive-Covid-19-Indian-zoo.html (Accessed 06/01/23).

49 Independent: Hong Kong says controversial hamster cull is complete and pet shops can resume sales: https://www.independent.co.uk/asia/china/hong-kong -hamster-culling-covid-b2004087.html (Accessed 06/01/23).

50 Mail online: Chinese city's plan to kill all pets belonging to Covid-19 patients in coronavirus hotspot neighbourhood is axed following outcry: https:// www.dailymail.co.uk/news/article-10671925/Chinese-citys-plan-KILL -pets-belonging-Covid-19-patients-axed-following-outcry.html (Accessed 06/01/23).

51 BBC: Vietnam: Owners heartbroken after 12 dogs killed over Covid: Vietnam: Owners heartbroken after 12 dogs killed over Covid – BBC News (Accessed 14/04/23).

52 Guardian: Millions of farm animals culled as US food supply chain chokes up: https://www.theguardian.com/environment/2020/apr/29/millions-of-farm

-animals-culled-as-us-food-supply-chain-chokes-up-coronavirus (Accessed 09/01/23).

53 World Nuclear Association: Chernobyl Accident 1986: https://world-nuclear.org /information-library/safety-and-security/safety-of-plants/chernobyl-accident .aspx (Accessed 06/01/23).

54 Wales online: How Chernobyl made Welsh sheep radioactive and paralysed some farms for 26 years: https://www.walesonline.co.uk/news/wales-news/how-cher-nobyl-made-welsh-sheep-16360676 (Accessed 06/01/23).

55 BBC: Fukushima disaster: What happened at the nuclear plant? https://www.bbc .co.uk/news/world-asia-56252695 (Accessed 06/01/23).

56 Wired: The Chernobyl disaster may have also built a paradise: https://www.wired .com/story/the-chernobyl-disaster-might-have-also-built-a-paradise/ (Accessed 09/01/23).

57 Chernobyl Guide: Chernobyl catfish and mutant fish – human's water friends: https://chernobylguide.com/chernobyl_catfish/ (Accessed 09/01/23).

58 Møller, A.P. and Mousseau, T.A. (2007) Species richness and abundance of forest birds in relation to radiation at Chernobyl. *Biology Letters*, *3*, 483–486.

59 NPR: Humans traveled less during COVID restrictions. Animals traveled more: https://www.npr.org/2023/06/08/1181043524/humans-traveled-less-during -covid-restrictions-animals-traveled-more?utm_source=Nature+Briefing&utm _campaign=e63a17c305-briefing-dy-20230609&utm_medium=email&utm_t erm=0_c9dfd39373-e63a17c305-45565222 (Accessed 21/07/23).

60 Guardian: Israel's ibex make the most of lockdown – in pictures: https://www .theguardian.com/world/gallery/2021/jan/22/israels-ibex-make-the-most-of -lockdown-in-pictures (Accessed 09/01/23).

61 Country Living: 13 photos of animals taking over towns and cities on lockdown: https://www.countryliving.com/uk/news/g32066174/animals-deserted-towns -cities-lockdown/ (Accessed 09/01/23).

62 RFI: Wild animals wander through deserted cities under Covid-19 lockdown: https://www.rfi.fr/en/international/20200330-wild-animals-wander-through -deserted-cities-under-covid-19-lockdown-ducks-paris-puma-santiago-civet -kerala (Accessed 09/01/23).

63 Bristol City Council: Urban foxes: https://www.bristol.gov.uk/residents/pests -pollution-noise-and-food/pest-control/urban-foxes (Accessed 09/01/23).

64 BBC: Coronavirus: Animals take over cities during self-isolation: https://www .bbc.co.uk/newsround/51977924 (Accessed 09/01/23).

65 BBC: Coronavirus: Wild animals enjoy freedom of a quieter world: https://www .bbc.co.uk/news/world-52459487 (Accessed 09/01/23).

66 Yahoo News: Coronavirus: Emboldened animals reclaim city streets as millions stay indoors in lockdown: https://uk.news.yahoo.com/coronavirus-animals -reclaiming-cities-covid-19-pandemic-140503907.html?guccounter=1&guce_ referrer=aHR0cHM6Ly93d3cuZWNvc2lhLm9yZy8&guce_referrer_sig=AQ AAAIhxlhl2NWBf1soMTHKzRYQqbGIwYFMl2c9KCUlZlhYrzM6auHZ _NvkDXpg8wdm8rzVustM9JRjNnhGntVD7qJP_FaXAD0hN-4w4nHa4aZAF JcBlT3DncVPZhGkOF0r7MBdP6bhr2gSql5tNPa41PCz531qEvkQ2TBO pRJDQDusA (Accessed 09/01/23).

67 Timeout: The animals taking back human spaces while the world is in lockdown: https://www.timeout.com/things-to-do/animals-taking-back-human-spaces -while-the-world-is-in-lockdown (Accessed 09/01/23).

68 Plataforma SINC. (2015) Studying scavenge hunting animals remaining world-wide. *ScienceDaily*: www.sciencedaily.com/releases/2015/08/150804074045.htm (Accessed 09/01/23).

69 CNBCTV: Coronavirus lockdown: As stray animals go without food and water, you could help make a difference: https://www.cnbctv18.com/healthcare/coronavirus-lockdown-as-stray-animals-go-without-food-and-water-you-could-help-make-a-difference-5674351.htm (Accessed 09/01/23).

70 Source: National Audit Office report 25th Nov 2020: https://www.nao.org.uk/reports/supplying-the-nhs-and-adult-social-care-sector-with-personal-protective-equipment-ppe/#conclusion (Accessed 14/08/23).

71 The Hill: States spent $7B in spring on medical devices, PPE: Report: https://thehill.com/homenews/state-watch/530985-states-spent-7-billion-in-spring-on-medical-devices-ppe-report/ (Accessed 09/01/23).

72 WMW: PPE waste: The problem with throwaway protection: https://waste-management-world.com/resource-use/ppe-waste-the-problem-with-throwaway-protection/ (Accessed 09/01/23).

73 BBC: Covid: Disposable masks pose pollutants risk, study finds: https://www.bbc.co.uk/news/uk-wales-56972074 (Accessed 09/01/23).

74 Pandey, G. and Madhuri, S. (2014) Heavy metals causing toxicity in animals and fishes. *Research Journal of Animal, Veterinary and Fishery Sciences*, 2, 17–23.

75 Shah, F.U.R., Ahmad, N., Masood, K.R. and Peralta-Videa, J.R. (2010) Heavy metal toxicity in plants. In *Plant Adaptation and Phytoremediation* (pp. 71–97). Springer, Dordrecht.

76 Ghori, N.H., Ghori, T., Hayat, M.Q., Imadi, S.R., Gul, A., Altay, V. and Ozturk, M. (2019) Heavy metal stress and responses in plants. *International Journal of Environmental Science and Technology*, 16, 1807–1828.

77 Wang, C., Zhao, J. and Xing, B. (2021) Environmental source, fate, and toxicity of microplastics. *Journal of Hazardous Materials*, 407, 124357.

78 Vivekanand, A.C., Mohapatra, S. and Tyagi, V.K. (2021) Microplastics in aquatic environment: Challenges and perspectives. *Chemosphere*, 282, 131151.

79 Roy, T., Dey, T.K. and Jamal, M. (2023) Microplastic/nanoplastic toxicity in plants: An imminent concern. *Environmental Monitoring and Assessment*, 195, 1–35.

80 Fendall, L.S. and Sewell, M.A. (2009) Contributing to marine pollution by washing your face: Microplastics in facial cleansers. *Marine Pollution Bulletin*, 58, 1225–1228.

81 WWF: In the disposal of masks and gloves you need responsibility: https://www.wwf.it/pandanews/ambiente/nello-smaltimento-di-mascherine-e-guanti-serve-responsabilita/ (Accessed 09/01/23).

82 De-la-Torre, G.E., Rakib, M.R.J., Pizarro-Ortega, C.I. and Dioses-Salinas, D.C. (2021) Occurrence of personal protective equipment (PPE) associated with the COVID-19 pandemic along the coast of Lima, Peru. *Science of The Total Environment*, 774, 145774.

83 The Ecoexperts: How long does PPE take to degrade naturally? https://www.theecoexperts.co.uk/blog/ppe-degradable (Accessed 09/01/23).

84 De-la-Torre, G.E., Dioses-Salinas, D.C., Dobaradaran, S., Spitz, J., Keshtkar, M., Akhbarizadeh, R., Abedi, D. and Tavakolian, A. (2022) Physical and chemical degradation of littered personal protective equipment (PPE) under simulated environmental conditions. *Marine Pollution Bulletin*, 178, 113587.

85 Pizarro-Ortega, C.I., Dioses-Salinas, D.C., Severini, M.D.F., López, A.F., Rimondino, G.N., Benson, N.U., Dobaradaran, S. and De-la-Torre, G.E. (2022) Degradation of plastics associated with the COVID-19 pandemic. *Marine Pollution Bulletin*, 113474.

86 Yang, S., Cheng, Y., Liu, T., Huang, S., Yin, L., Pu, Y. and Liang, G. (2022) Impact of waste of COVID-19 protective equipment on the environment, animals and human health: A review. *Environmental Chemistry Letters*, 20, 2951–2970.

87 Hiemstra, A.F., Rambonnet, L., Gravendeel, B. and Schilthuizen, M. (2021) The effects of COVID-19 litter on animal life. *Animal Biology*, *71*, 215–231.

88 Guardian: Crows and magpies using anti-bird spikes to build nests, researchers find: https://www.theguardian.com/science/2023/jul/11/crows-and-magpies-show-their-metal-by-using-anti-bird-spikes-to-build-nests (Accessed 12/07/23).

89 De-la-Torre (2021).

90 Guardian: Takeaway food and drink litter dominates ocean plastic, study shows: https://www.theguardian.com/environment/2021/jun/10/takeaway-food-and-drink-litter-dominates-ocean-plastic-study-shows (Accessed 17/02/23).

91 Litter, it costs you: 9 interesting facts and statistics about littering: https://litteritcostsyou.org/9-interesting-facts-and-statistics-about-littering/ (Accessed 17/02/23).

92 WHO: Modes of transmission of virus causing COVID-19: Implications for IPC precaution recommendations: https://www.who.int/news-room/commentaries/detail/modes-of-transmission-of-virus-causing-covid-19-implications-for-ipc-precaution-recommendations (Accessed 14/02/23).

93 Nature: COVID-19 rarely spreads through surfaces. So why are we still deep cleaning? https://www.nature.com/articles/d41586-021-00251-4#ref-CR1 (Accessed 14/12/23).

94 Van Doremalen, N., Bushmaker, T., Morris, D.H., Holbrook, M.G., Gamble, A., Williamson, B.N., Tamin, A., Harcourt, J.L., Thornburg, N.J., Gerber, S.I. and Lloyd-Smith, J.O. (2020) Aerosol and surface stability of SARS-CoV-2 as compared with SARS-CoV-1. *New England Journal of Medicine*, *382*, 1564–1567.

95 Medical News Today: How long does the virus last on clothes and surfaces? https://www.medicalnewstoday.com/articles/how-long-does-coronavirus-last#on-surfaces-and-clothes (Accessed 14/02/23).

96 Van Doremalen, N., Bushmaker, T., Morris, D.H., Holbrook, M.G., Gamble, A., Williamson, B.N., Tamin, A., Harcourt, J.L., Thornburg, N.J., Gerber, S.I. and Lloyd-Smith, J.O. (2020) Aerosol and surface stability of SARS-CoV-2 as compared with SARS-CoV-1. *New England Journal of Medicine*, *382*, 1564–1567.

97 Corpet, D.E. (2021) Why does SARS-CoV-2 survive longer on plastic than on paper? *Medical Hypotheses*, *146*, 110429.

98 Euronews: Omicron survives much longer on plastic and skin than earlier COVID variants, new study finds: https://www.euronews.com/next/2022/01/24/omicron-survives-much-longer-on-plastic-and-skin-than-earlier-covid-variants-new-study-fin (Accessed 18/07/23).

99 BBC: Coronavirus: How to wash your hands – in 20 seconds: https://www.bbc.co.uk/news/av/health-51754472 (Accessed 14/02/23).

100 The Atlantic: Large-scale disinfection efforts against Coronavirus: https://www.theatlantic.com/photo/2020/03/photos-large-scale-disinfection-efforts-against-coronavirus/607810/ (Accessed 14/02/23).

101 U.S. Environmental Protection Agency: Can I use fogging, fumigation, or electrostatic spraying, or drones to help control COVID-19?: https://www.epa.gov/coronavirus/can-i-use-fogging-fumigation-or-electrostatic-spraying-or-drones-help-control-covid-19 (Accessed 14/02/23).

102 Guardian: 10 Covid-busting designs: Spraying drones, fever helmets and anti-virus snoods: https://www.theguardian.com/artanddesign/2020/mar/25/10-coronavirus-covid-busting-designs (Accessed 14/02/23).

103 Mohsan, S.A.H., Zahra, Q.U.A., Khan, M.A., Alsharif, M.H., Elhaty, I.A. and Jahid, A. (2022) Role of drone technology helping in alleviating the COVID-19 pandemic. *Micromachines*, *13*, 1593.

104 MSN: Thames Water among worst water companies in country due to pollution, damning report says: https://www.msn.com/en-gb/money/other/thames-water -among-worst-water-companies-in-country-due-to-pollution-damning-report -says/ar-AA1dYuW0 (Accessed 18/07/23).

105 Bhat, S.A., Sher, F., Kumar, R., Karahmet, E., Haq, S.A.U., Zafar, A. and Lima, E.C. (2022) Environmental and health impacts of spraying COVID-19 disinfectants with associated challenges. *Environmental Science and Pollution Research International*, *29*, 85648–85657.

106 National Post: How Canada's addiction to road salt is ruining everything: Bringing down bridges, melting cars, poisoning rivers; it's hard to think of something salt isn't ruining: https://nationalpost.com/news/canada/how-canadas-addiction-to -road-salt-is-ruining-everything (Accessed 31/01/23).

107 Government of Canada: Road salts: Frequently asked questions: https://www .canada.ca/en/environment-climate-change/services/pollutants/road-salts/fre- quently-asked-questions.html (Accessed 14/02/23).

108 Deeproot: Deicing with beet juice: https://www.deeproot.com/blog/blog -entries/deicing-with-beet-juice/ (Accessed 17/02/23).

109 EcoMyths: Beet this: Morton Arboretum's clever snow removal ingredient: https://ecomyths.org/beet-this-morton-arboretums-clever-snow-removal -ingredient/ (Accessed 17/02/23).

110 Zhongming, Z., Linong, L., Xiaona, Y. and Wei, L. (2018) Plant-based 'road salt' good for highways but not for insects: https://www.sciencedaily.com/releases /2018/10/181029135232.htm (Accessed 17/02/23).

111 Ho, J., Hussain, S. and Sparagano, O. (2021) Did the COVID-19 pandemic spark a public interest in pet adoption? *Frontiers in Veterinary Science*, 444.

112 The Dog People: The pandemic pet adoption boom: What we've learned, one year later: https://www.rover.com/uk/blog/pandemic-pet-adoption-boom/ (Accessed 17/02/03).

113 Morgan, L., Protopopova, A., Birkler, R.I.D., Itin-Shwartz, B., Sutton, G.A., Gamliel, A., Yakobson, B. and Raz, T. (2020) Human–dog relationships during the COVID-19 pandemic: Booming dog adoption during social isolation. *Humanities and Social Sciences Communications*, *7*(1).

114 Guardian: Pets prove to be the pandemic's cute, furry growth area: https://www .theguardian.com/news/datablog/2022/jan/21/pets-ownership-pandemic-dogs -cats (Accessed 17/02/23).

115 Mueller, M.K., Richer, A.M., Callina, K.S. and Charmaraman, L. (2021) Companion animal relationships and adolescent loneliness during COVID-19. *Animals*, *11*, 885.

116 Grajfoner, D., Ke, G.N. and Wong, R.M.M. (2021) The effect of pets on human mental health and wellbeing during COVID-19 lockdown in Malaysia. *Animals*, *11*, 2689.

117 Tan, J.S.Q., Fung, W., Tan, B.S.W., Low, J.Y., Syn, N.L., Goh, Y.X. and Pang, J. (2021) Association between pet ownership and physical activity and mental health during the COVID-19 "circuit breaker" in Singapore. *One Health*, *13*, 100343.

118 Phillipou, A., Tan, E.J., Toh, W.L., Van Rheenen, T.E., Meyer, D., Neill, E., Sumner, P.J. and Rossell, S.L. (2021) Pet ownership and mental health during COVID-19 lockdown. *Australian Veterinary Journal*, *99*, 423–426.

119 Barklam, E.B. and Felisberti, F.M. (2022) Pet ownership and wellbeing during the COVID-19 pandemic: The importance of resilience and attachment to pets. *Anthrozoös*, 1–22.

120 Psychology Today: What you didn't know about having a pandemic pet: https://www.psychologytoday.com/us/blog/animals-and-us/202205/what-you-didnt-know-about-having-a-pandemic-pet (Accessed 17/02/23).

121 Huang, Q., Zhan, X. and Zeng, X.T. (2020) COVID-19 pandemic: Stop panic abandonment of household pets. *Journal of Travel Medicine*, 27, taaa046.

122 As we write this (February 2023) the war is still going on.

123 BBC: Cost of living: More pets being abandoned, dogs home says: https://www.bbc.co.uk/news/uk-wales-62610727 (Accessed 18/02/23).

124 RSPCA: Meet the unwanted lockdown puppies: https://www.rspca.org.uk/-/blog-meet-the-unwanted-lockdown-puppies (Accessed 18/02/23).

125 BBC: Covid: Dogs bought in lockdown being abandoned: https://www.bbc.co.uk/news/uk-wales-58996017 (Accessed 18/02/23).

126 BBC: Covid: Concerns over 'dogfishing' and abandoned pets: https://www.bbc.co.uk/news/uk-wales-54643823 (Accessed 18/02/23).

127 The Daily Mail: 'Lockdown puppies' flood rescue centres: Hundreds of pets are being abandoned as owners who bought them for company during pandemic struggle to cope with caring for them: https://www.dailymail.co.uk/news/article-9109495/Hundreds-pets-abandoned-owners-bought-pandemic-struggle-cope.html (Accessed 18/02/23).

128 Irish Examiner: Ireland in midst of 'pig crisis' as pandemic pets now abandoned: https://www.irishexaminer.com/news/arid-41057711.html (Accessed 18/02/23).

129 Daily Paws: Millions of pets weren't spayed and neutered during the pandemic and that's a big problem: https://finance.yahoo.com/news/millions-pets-werent-spayed-neutered-223127905.html (Accessed 18/02/23).

130 Guerios, S.D., Porcher, T.R., Clemmer, G., Denagamage, T. and Levy, J.K. (2022) COVID-19 associated reduction in elective spay-neuter surgeries for dogs and cats. *Frontiers in Veterinary Science*, 9.

131 van Dobbenburgh, R. and De Briyne, N. (2020) Impact of Covid-19 on animal welfare. *The Veterinary Record*, 187, e31–e31.

132 Guardian: Want to truly have empathy for animals? Stop owning pets: https://www.theguardian.com/commentisfree/2023/feb/04/want-to-truly-have-empathy-for-animals-stop-owning-pets (Accessed 18/02/23).

133 Organisms not containing a nucleus or organelles, such as bacteria.

7

Animals, pandemics, and the future

7.1 Introduction

According to the WHO, between 2011 and 2017 there have been 1,307 epidemic events across 172 countries. Astonishingly this averages an epidemic event every two days.[1] The WHO also says that any pandemic can be broken down into five stages:

■ Anticipation.
■ Early detection.
■ Containment.
■ Control and mitigation.
■ Eradication.

The same report says that "if human beings want to be healthy, we need a healthy environment with healthy animals." It also says that "70% of emerging human pathogens come from animals." Therefore, considering animals before and during an epidemic/pandemic is very important. Having good animal management can help in all five stages of a pandemic, from anticipation (can we predict where the next virus will emerge and from what species? – probably a bat, although influenza was thought to originate in fish[2]) through containment (stopping animals spreading the disease) to eradication (stopping animals becoming hosts for possible mutations and reemergence of new viral variants).

In a *60 Minutes* interview for CBS in September 2022, the president of the United States declared that the COVID-19 pandemic was over.[3] This

DOI: 10.1201/9781003427254-7

may have been technically correct, as most countries were trying to get back to normal, flights had resumed in most places, trains were operating normally, and very few people were wearing masks or social distancing. In the United Kingdom, as of December 2022, it is rare to see people wearing masks in supermarkets and on trains. It was not long ago that we were forced to social distance outside of supermarkets and had a one-in/one-out policy. So, in many respects the US president may have been correct. However, others argue that this is not the case. Dr Peter Hotez[4] warns that President Biden is wrong and that too many lives are still being lost.[5] Furthermore, a look at the news emanating from China at the moment (end of 2022/beginning of 2023), shows that the COVID-19 infections are not to be ignored. In December 2022, Qingdao in China was reporting between 490,000 and 530,000 new COVID-19 cases a day in a city of about 10 million people.[6] It appears that these numbers were quickly removed from subsequent reports, so this is hard to substantiate. On the other hand, other media outlets are reporting similar news. *US News* stated that 248 million people were infected across China in the first 20 days of December 2022, with 37 million thought to be infected in one day,[7] a report supported by an article in *Bloomberg UK* on 23 December 2022.[8] These are hardly reports that support the notion that the pandemic is over, as President Biden seemed to think.

It is becoming obvious now that COVID-19 will not be going away for the foreseeable future, if ever. Like the common cold and influenza, it will be here and causing problems for a long time. There is, for a significant number of people, little risk with allowing the virus to continue. What are the alternatives? on-and-off lockdowns and the impact of those, including reduced mental health and economic depression. But on the other hand, there are also a significant number of people who are likely to be severely affected by the SARS-CoV-2 virus. Many populations are still not fully vaccinated, and there are numerous vulnerable groups, including the elderly and those with underlying medical conditions. Therefore, we cannot be too complacent. And then there are the animals, which we still know little about.

It is with this in mind that we need to look to the future. What will the COVID-19 pandemic continue to bring, and what should we do about it? And what can we learn so when it happens again, we are better prepared? In January 2023 there were news reports of a new Omicron variant. A variant labelled as XBB was seen in the United Kingdom in September 2022, but there is now an evolution of this called XBB.1.5, and this is causing some concern.[9] Such reports highlight the uncertainty of the future when considering COVID-19.

However, it is not all doom and gloom. For example, on 16 May 2023, the newspapers were reporting that the UK had announced a plan for a "genomics transformation," so that future pandemics could be prepared for. According to the *Guardian*, "The UK Health Security Agency (UKHSA) now aims to build on the lessons of the pandemic by embedding genomics into routine public health practice."[10] Interestingly there was little mention of animals, except as a source of future viruses. Bird flu was given as an example of the power of genomics, but mainly as it helps to understand how viruses may jump to mammals, with the implication that humans are at risk. Certainly, any expansion of genomics and information could be expanded to protecting species other than humans.

7.2 Long COVID – what we know: a concern for animals?

One of the problems of having been infected with SARS-CoV-2 is that for some people there are long-lasting symptoms. This is usually called long Covid, but is otherwise referred to as post COVID-19 syndrome. The symptoms that have been reported are listed in Table 7.1.

However, recently a list of the 12 most likely long Covid symptoms were published (narrowed down from over 200), and these include changes in the sense of smell, post-exertional malaise, chronic cough, brain fog, and thirst. The study was based on self-reporting, and there might have been some selection bias of participants,[12] but it was deemed to be a significant step to understanding the disease. After all, many of the long-term effects of COVID-19 can be life debilitating, and they should not be dismissed. There is not only the welfare cost for individuals, but also a significant financial cost to medical services. It has been estimated that the average medical cost in the first six months of long COVID may be $9,500,[13] so not insignificant. Therefore, even if the COVID-19 pandemic finally ends, and the virus stops spreading through the human population, the effects it leaves will reside for years.

Long COVID is referred to as a human syndrome, and there has been little or nothing said about it in animal populations. Even though animals have died through a SARS-CoV-2 infection and many of the acute symptoms are the same, such as coughing and the symptoms of flu, there is little known about whether long-term effects are likely in animals. Why should this be reserved to the human population only? There is no molecular biological reason for long COVID being restricted to humans. After all, gorillas, which we know can become infected with the virus, are around 98% similar to humans at the genetic level.[14]

And this is, in fact, what is starting to be reported. In macaques there is some evidence that the animals have a long-term chronic response to a SARS-CoV-2 infection.[15]

Table 7.1 Some of the long-term effects in humans of having been infected with SARS-CoV-2, as listed on the UK Government website[11]

Effects of long Covid reported

Most common symptoms	Extreme tiredness (fatigue)
	Shortness of breath
	Loss of smell
	Muscle aches
Other symptoms reported	Problems with your memory and concentration ("brain fog")
	Chest pain or tightness
	Difficulty sleeping (insomnia)
	Heart palpitations
	Dizziness
	Pins and needles
	Joint pain
	Depression and anxiety
	Tinnitus, earaches
	Feeling sick, diarrhoea, stomach aches, loss of appetite
	High temperature, cough, headaches, sore throat, changes to sense of smell or taste
	Rashes

It is not just the direct effects of long COVID on the animal's health that we need to be concerned about either. There is a worry that the relationship between a long-suffering animal and their carers/owners may be strained,[16] with the possible detriment to the welfare of the animal.

Long COVID is going to be an ongoing topic of discussion for human health as we travel into the future, but perhaps we should give greater consideration to the animals around us in this discussion. Many of the symptoms seen in humans, and experienced by people trying to continue their normal lives, are likely to have a significant impact on animal welfare, if translated to an animal population. Symptoms such as fatigue, joint pain, dizziness, and tinnitus will not be good for any animal. In captivity, or companion animals, such symptoms may be able to be diagnosed, and even treatments used to mitigate the issue, but what about those animals in the wild? Even if not a widespread problem, it is likely that individual animals will suffer the lasting effects of a SARS-CoV-2 infection for many months, or even years, and most of this animal suffering may well go unnoticed. Perhaps we need to be more cognisant of this as pandemics such as COVID-19 spread across the globe in the future.

7.3 What can be learnt from COVID-19 for the next epidemic/pandemic

It is likely that there will be another pandemic during our lifetime, says a report from the BBC,[17] although the definition of "our lifetime" is a little vague. But even taking a longer-term view here, it seems almost inevitable. So what can we learn from COVID-19 that may prepare us for this next event?

A wide range of emergent pathogens was reviewed recently by Reperant and Osterhaus.[18] They list several in the title of their article: "AIDS, Avian flu, SARS, MERS, Ebola, Zika... what next?" Rather depressingly they say:

> Due to the complex and largely interactive nature of the drivers for newly emerging virus infections, it is virtually impossible to predict what the next pathogen threat will be, from where it will come and when it will strike.

However, this does not mean we should not try to be prepared.

Nature magazine ran an article by Devi Sridhar on 26 October 2022[19] listing five issues that ought to be considered so we are more prepared for the next time. These were:

- More monitoring of potential zoonotic diseases. It says that WHO has identified several diseases with pandemic potential, including Ebola, SARS-like diseases, Lassa fever, and Zika. Sridhar says that there should be more vigilance where animals and humans have close contact, and better hygiene in such places, including markets where animal products are sold.
- More sequencing capacity. Although many of the SARS-CoV-2 variants were sequenced very rapidly once isolated, the capacity to achieve this is patchy across the globe. It is interesting to note that this led to a misconception (amongst many politicians) that certain countries were "hotbeds" of variants, e.g., the United Kingdom and South Africa. This was simply due to these countries having much more highly developed genomic sequencing capacity than

most others. In fact, the United Kingdom accounted for 38% of all SARS-CoV-2 sequencing.[20]

■ More manufacturing capacity. Here the focus is on the production of vaccines and therapeutics. Again, whilst some countries have superb facilities, the picture across the globe is not consistent. During the COVID-19 pandemic there was a call to share vaccines, for example, but this was not widely put into practice. Gordon Brown, former prime minister of the United Kingdom, said that such a lack of actions was a "stain on our global soul."[21]

■ It was soon found during the COVID-19 pandemic that the only true way to stop the virus spread and keep people healthy was to have an effective vaccine. Sridhar suggests that the stockpiles of suitable vaccines, when possible, should be prepared in advance, so there is less delay in having populations vaccinated. She suggests that technology needs to be ready to adapt known vaccines, in what she describes as plug and play so that new variants can be produced quickly off known and existing platforms. We would suggest that such thoughts be also extended to animal populations, especially for any species that may be involved in future pandemic spread, such as companion animals.

■ Social measures to stop the spread of any virus need to be accepted and put in place quickly, and this together with an efficient vaccine roll out should help human, and therefore by default animal, health. Mask wearing needs to be made mandatory and safe space should be identified so that essential services, such as education, are not curtailed as they were during the COVID-19 pandemic. In Denmark school classes were moved to football stadia, allowing classes to continue in a socially distanced manner,[22] the type of measure that could easily be adopted, and indeed planned for, during the next pandemic.

These measures were presented to the G7[23] as part of a 100 Days Mission. Whether this will be enacted is yet to be seen.

With a focus on primates, others have also come up with a list of recommendations for the future. These include having better biosecurity and to model safe spaces where human-primate interactions can make essential actions less likely to spread any future virus, having better data collection about how human and primate movements can affect each other, minimising the provisioning or capture of wild animals, and having more focused funding to address issues with human-primate interactions. For the long term there was a call for better collaboration between interested parties, better health monitoring, and better education focused on this topic.[24]

However, many of these ideas are not that new. As the Ebola epidemic was declining during the summer of 2015, Ross and colleagues[25] were calling for future planning, which included:

■ Meeting the challenges in abiding by international health regulations, including strengthening global health systems.
■ Putting in place global pandemic funding.
■ Taking a One Health approach for future pandemic planning, which, of course, would include considering animals and wider ecosystems in the conversation.

Of course, such a discussion should not be focused on COVID-19 and SARS-like viruses. There has been much discussion in the past about how to deal with an influenza epidemic, and such ideas and knowledge can be translated across to any epidemic/pandemic. In 2003, with a review of influenza epidemics, there was a call that "contingency plans must be put in place now during the inter-pandemic period."[26] In 2006 WHO published a document entitled "Global pandemic influenza action plan to increase vaccine supply."[27]

Others have focused on the ethical issues which will be raised in both preparing for a pandemic and during the event itself. Rather alarmingly, a report published in 2005, looking at the forward planning of Canada, the United Kingdom, and the United States, stated that plans "do not identify resources that would be optimally required to reduce deaths and other serious consequences," and they say that any future plans should lead to "an atmosphere of mutual trust and solidarity."[28] Another paper suggests that back in 2015 pandemic plans were "ill-equipped to anticipate and facilitate the navigation of unique ethical challenges"[29] that would arise.

Animals are not excluded from discussions around future planning. In a paper by Taubenberger in 2020, it was suggested that future planning needs to have better surveillance, which "also needs to include humans, domestic animals, and wild birds."[30] Others agree, with the COVID-19 pandemic in full swing in 2021, Thoradeniya and Jayasinghe[31] suggest that there are five steps that need to be considered when thinking about any future pandemic, and animals are integral to their thinking. These five steps are listed as:

1. Origins in the animal kingdom; 2. Transmission to domesticated animals; 3. Inter-species transmission to humans; 4. Local epidemics; 5. Global spread towards a pandemic.

And they go on to say: "A global view using a systems science approach is necessary to recognize the close interactions between health of animals, humans and the environment."

There is, therefore, a body of discussion, which has been around long before SARS-CoV-2 emerged, that looks at how countries and the world as a whole could, or indeed should, plan for a future pandemic. Looking back at the early days of COVID-19 we saw quite a lot of panic, a lot of rushing to put procedures in place and to secure equipment. There seemed to be little evidence of pragmatic forward planning, and indeed during the United Kingdom's COVID inquiry it was recognised that planning had been inadequate.[32] Had animals been considered? One can only speculate that the answer was that if they were considered, it was not enough to stop the emergence of a truly significant virus. It almost certainly came from a bat, so clearly biosecurity was not adequate. The virus may have "jumped" from the bat to humans via at least one other animal, and therefore surveillance, monitoring and better hygiene – all things called for by suggested pandemic planning – was not put in place. It is easy to be critical with hindsight, but we should not forget that we had already had SARS and MERS.

Many of the pandemic plans need to be re-visited and taken more seriously by many governments, not just those where a future outbreak is possible, or indeed likely. The United Kingdom and many other countries were caught without a plan that could be quickly executed – indeed the UK Government delayed doing much to start with,[33] with some accusing the UK Government of wanting the virus to spread. However, there were calls for a better preparedness and a better monitoring and surveillance of animals. Could COVID-19 have been prevented? We will never know, but it looks certain that we need to plan well for the next pandemic.

7.3.1 Should there be more focus on animals and why?

Well into the third year after the realisation that there was a COVD-19 outbreak, there is still a call for animals to be monitored for coronaviruses.[34] Many species harbour this class of virus, and there are concerns that some will mutate and then jump to humans.[35] New forms are less likely to be sensed well by the human immune system and can lead to further viral outbreaks. After all, this is the likely route by which COVID-19 started in the first place, and it is very likely to be repeated with novel viruses (or viral strains) in the future.

Crary and Gruen[36] have postulated that ecofeminism is a useful lens through which to view the future of animal-human relations, stating: "This framework enables us to see practices that destroy nature, animals and marginalised human groups as structurally interrelated." In this context, it is easy to see that both animals and marginalised humans were not given the same priority as non-marginalised citizens of wealthy countries.

Marginalised humans were not able to get vaccines (and in that context it is inconceivable that animals should be prioritised for a vaccine), or health care, or even burial services. Animals were culled. A One Health, One Medicine approach, where prevention (including, for example the use of vaccines, and not catching wild animals and keeping them alive in wet markets), health care and services are provided in an emergency situation not just by where you live in the world, whether you are part of an elite group, but on the basis of your need as a human, or a non-human, sentient living creature (such as the One Medicine approach we saw of making modified human vaccines available to gorillas). We are all part of an ecosystem, and the pandemic writ large showed how the health of animals is fundamental to the health of humans, and that human (mis-) treatment of animals may have led to the catastrophe that followed the discovery of the virus in a wet market in Wuhan. There has to be more of a focus on animals, for all our sakes.

Much of the discussion about why animals are important for the consideration of future pandemics has already been covered. But it is worth pulling together a few salient points here:

- There will be future pandemics, and they will almost certainly be caused by a virus that emerges from an animal, very possibly a bat. Bats are huge reservoirs of coronaviruses, and they have been the source of past SARS-like viruses. Future viruses might emanate from bats too, and not just in the Far East. In June 2023 a paper in *Nature* highlighted that bats in the United Kingdom were reservoirs for SARS-like viruses,[37] so it likely to be a global issue. Therefore, better biosecurity is needed here. This does not mean mass cullings, but it may mean a prohibition on eating them.
- Bat-to-human transmission seems to require another animal species. This could be camelids, or something else. The intermediate in COVID-19 has still not definitively been identified. Therefore, handling of animals and products from dead animals needs to be taken very seriously.
- We need to be vigilant of viral sources that may not strike people as obvious. Bird flu? Who cares? It is flu in birds (although it has certainly resulted in some human deaths) – but when it spreads to humans it may be too late to put the genie back in the bottle. Good communication and education around such issues are needed to bring our interaction with wild animals into the fore of people's thinking.
- Animals die because of viruses spreading through the human population. In COVID-19 this was not as widespread as it could have been, but many animals of different species did die. No doubt there were larger numbers of deaths than we will ever know.

- Animals may spread the virus. We do not know how widespread this was, or indeed still is, in the COVID-19 pandemic, as sufficient monitoring and data collection were never carried out. It is possible that, in future pandemics, domestic or wild animals may prove to be significant vectors for transmission. Bird flu can be contracted by mammals, and if this moves from predation based to mammal-to-mammal transfer, then we could face severe population pressure, if not in some cases collapse, for some mammal species. In addition to the glaringly obvious catastrophe of such a high mortality disease in humans, in a circumstance where potentially all humans, and all mammals domestic or in the natural environment may be possibly infectious, it does not take much of a leap of imagination to envision the scale of potential animal culling operations that might be set in place.
- Animals are good reservoirs of viruses and as a virus passes through an animal it can mutate. Many new viral variants can emerge from animals, and we need to be vigilant about this. It seems an opportunity lost if mink farms are allowed to be re-established in the same manner as they were before COVID-19, and other animal management practices need to be looked at too.
- Animals, whether you agree or not in relation to their use, are good "models" for the study of virus activity. This is unlikely to stop in the near future, as it is considered that a lot can be learnt from how a virus invades a whole animal and how a disease progresses. Much of this can be done in cell culture, but there will no doubt be seen that there is a use for animal models for the foreseeable future (animal models for COVID-19 are further discussed in Section 4.4).
- And then there are all the indirect effects on animals that a pandemic brings, from entanglement in PPE to more poaching and less funding for conservation.

We only have one lump of rock we all share, and animals are an integral part of our ecosystem. We seem to often forget this, but we will revisit this theme in Section 7.7.

7.3.2 How can animal welfare be preserved and protected in a pandemic?

We need a less human-centric approach to pandemic management. There is an unfortunate propensity to blame the animals for inflicting the pandemic on us. It was not the bats' fault. It was humans who made the mistake, not the bats, which were simply going about their normal lives. Again, much of this has been said before, but let us sum up some issues again here:

- Animals need not be persecuted because they were the source of a pandemic-causing virus. Bats and pangolins need not be killed.
- Let us have less PPE lying about in the environment.
- Let us use fewer disinfectants and toxic chemicals in what appears to be a blanket war against the virus. All these chemicals have to end up somewhere, often in the rivers and the seas. It seems as though alternate methods of disinfecting materials need to be developed and adopted.
- Let us preserve funding for conservation and animal security.
- Let us have more monitoring of animal health and virus infection.
- Let us learn that keeping animals in tight cages is not only bad for animal welfare, but can ultimately be bad for human health too, as viruses can emerge from such places.

And finally, let us not forget that many animals, including ones yet to be considered so, are sentient beings. They have feelings, they hurt, and they cry. They are not that different from us in that regard, and we often seem to forget this. We should learn to treat animals as we would like to be treated ourselves and not consider them as lesser organisms to which we can do what we like. Such attitudes bode poorly for animal welfare, and, moreover, they are likely to come back and bite us when the next pandemic arrives.

7.3.3 Use of animals in COVID-19 detection

The extraordinary olfactory abilities of domestic dogs have long been recognised and, over the last century, dogs have increasingly been used in search and rescue activities (e.g., for earthquake/avalanche victims) and the detection of illicit substances at national borders (mostly narcotics and explosives). In recent years, however, there has been an increased interest in the use of these "sniffer" dogs to detect volatile compounds released by humans in certain diseases, including several different cancers,[38] neurodegenerative conditions,[39] and diabetes.[40] Amazingly, some dogs can be trained to detect compounds released by epilepsy sufferers up to 50 minutes before a seizure,[41] allowing the patient to move to a safe environment. Due to the high sensitivity and accuracy displayed by the dog olfactory system, there has been great interest in training sniffer dogs to detect COVID-19 in patients.

Such a detection system has several obvious advantages:

- It is well established that COVID-19 is asymptomatic in the first few days after infection, yet sufferers are still infectious during this time. This is one of the primary reasons for the rapid global spread of the disease, as asymptomatic individuals are unlikely to test for infection.
- The vast majority of tests currently performed for COVID-19 are either polymerase chain reaction (PCR)–based, or lateral-flow tests (LFTs). Whilst the former are highly sensitive, PCR tests require a fully equipped laboratory and at least 24 hours to generate a result. LFTs, on the other hand, can give a result in 15 minutes, but they are far less sensitive, especially in asymptomatic individuals. Such limitations make mass, accurate screening of individuals (e.g., at national borders or large social events) impractical.
- Use of COVID-19 sniffer dogs is essentially passive and non-invasive for the individuals being screened and, therefore, does not require the overt consent of the individuals being tested.
- Sniffer dogs are a practical solution in situations where access to high-tech screening devices and laboratories is limited.

WHO acknowledges that one of the main limitations in preventing the spread of infectious diseases is the ability to reliably screen large numbers of people accurately. One study suggested that COVID-19 sniffer dogs have an effective throughput of 250 humans an hour[42] or one test roughly every 14 seconds, an order of magnitude faster than even the most rapid LFTs. Another study has demonstrated a sensitivity of over 85% and a specificity of over 95% in the detection of COVID-19,[43] putting the sniffer dogs on a par with current laboratory testing procedures and ahead of many of the reported figures for LFTs on asymptomatic patients. It is also important to note that sniffer dog training for COVID-19 is currently in its infancy, and with refinements to this training, it is highly likely that greater specificity and sensitivity will be achieved.

Another key advantage in the use of sniffer dogs to detect COVID-19 is that they do not require the presence of the individual to be tested. Volatile signature compounds are readily deposited on clothes and PPE, and studies have found that detection using these items by dogs is comparable to exposure to the individual themselves.[44]

One caveat in this narrative is that, at the time of writing, COVID-19 testing is no longer required for entry into the majority of countries or participation in most mass-gathering events. However, at the current time (early 2023) there is a massive resurgence of COVID-19 in China, due to the sudden easing of the "zero-COVID" policy in that country.

Such huge infection numbers greatly increase the chances of the emergence of new, more infectious or "vaccine-escape" variants of COVID-19, meaning that there may be a need for increased border controls or social restrictions in the future. Given the figures above, even a very large airport such as London Heathrow (>100,000 passengers per day) would need only a relatively small number of COVID-19 sniffer dogs to effectively screen every passenger, obviating the need for expensive travel testing and cumbersome documents, all of which add to the increased delays seen at border security since the COVID-19 pandemic.

One question that remains to be answered is whether dogs could be trained to recognise novel COVID-19 variants, thus directing these cases for detailed sequencing. It seems unlikely that subtle changes in the virus could be detected (e.g., those caused by single nucleotide changes); however, detection of large-scale changes in the virus may well be possible. This is particularly true if the variant causes a change in virus symptoms, e.g., see Table 7.2 below:

Table 7.2 Most common symptoms of 3 major COVID-19 variants[45]

Variant	Most common symptoms
Alpha	Loss of taste or smell
	Fever
	Cough
	Shortness of breath
	Fatigue
	Headache
	Sore Throat
	Nasal congestion
	Nausea, vomiting, and diarrhoea
Delta	Sore throat
	Headache
	Runny nose
	Fever
	GI problems/diarrhoea
Omicron	Sore throat
	Headache
	Runny nose
	Congestion
	Cough
	Lower back pain
	Fatigue

Given the differences in these symptoms (and, by extension, differences in the cell types infected), it is conceivable that these variants would result in differences in the volatile compounds released by infected people and therefore it may be possible for sniffer dogs to detect this.

An obvious question is, if such sniffer dogs are used on a large scale for this or future diseases, how will the welfare of these working dogs be safeguarded? In the United Kingdom, there are very strict animal welfare laws and a high standard of care is applied to working animals of all kinds, but this may not be the case everywhere. Thus, the mass use of dogs for detection of COVID-19 or other diseases may itself create animal welfare concerns that need addressing.

7.4 Sources of future viruses and their transmission, the role of animals and of human interactions with animals

It is very probable, as discussed, that an epidemic or pandemic will arise from animals in the wild. It is likely that at some point a coronavirus, perhaps SARS-like, will emerge from a bat, cross to another animal species, and then to humans. COVID will happen again, and we will be in a fight against COVID-30 or COVID-33 or whatever year it happens. It is not if, but when. However, it is not the only source of human pathogens.

Xiao and colleagues,[46] during the height of the COVID-19 pandemic, looked back at the events of the Wuhan "wet market." This was thought to be the epicentre of the SARS-CoV-2 spread. They point to the poor animal welfare at such sites, how the hygiene is bad, and warn against the rise of potential new pathogens. They also highlight that on 26 January 2020 China temporarily banned all wildlife trade, and then on 24 February 2020 authorities brought in a ban on eating and trading terrestrial wild animals for food. However, there are now "wet markets" reopening, and it will be interesting how different practices will be compared to before the COVID-19 pandemic. If nothing changes, another virus emergence is likely to be just around the corner.

Humans are regularly in close contact with animals, and they always will be. We farm them, catch them, and eat them. We stroke and cuddle our pets. Close proximity to animals is inevitable. There are also times when our empathy forces us to be close to animals. Sick or injured animals tend to lead to human intervention. All these times that we are close to animals potentially may allow a human pathogen to also get close enough to take the opportunity to try to infect us. We do not know if an infected dog led to any humans catching COVID-19, but it certainly seems possible – a virus on the fur could easily to transferred to a person's eye, for example.

Such close proximity to animals is not a concern only with SARS-like infections. At the time of writing there are two possible concerns about the spread of disease from animals to humans: bird flu and monkeypox. These will be considered in the next sections of this text.

7.4.1 Avian influenza (avian flu, bird flu): should we be concerned?

A recent BBC radio show (9 February 2023) asked the question: How worried should we be about avian flu?[47] This is a good question. It may be that the virus will stay in the wild and all we will see is a few dead birds. Or it will spread to birds used for human consumption, such as chickens and ducks, and our food chains will be affected. Or it will spread to humans. Should we be worried?

Even if the disease just stayed in the wild it does not mean that humans will not get near the virus. Dead birds may be found in gardens or on beaches, as seen in Figure 7.1. Children, or even adults, may be tempted to play with such corpses. People may pick them up to move them, perhaps to bury them somewhere. Other animals will use the corpses as a food

Figure 7.1 A dead gannet on a beach, photographed by Ros Rouse© (but made black and white). The original image can be found at https://www .tigerloveart.co.uk.

source, and so be in contact and even eat the virus. So, should we be worried? A look at the Royal Society for the Protection of Birds (RSPB) web page is not cheerful reading, and it states that bird flu "has killed tens of thousands of birds" so far. The disease has affected at least 70 bird species in the United Kingdom alone, and they include:

> seabirds (Gannet, Great Skua, Common Guillemot, Atlantic Puffin, Razorbill, Black-legged Kittiwake, Northern Fulmar, gulls and terns), geese, ducks, swans and raptors. Our seabirds have been particularly badly hit.[48]

So, what is bird flu? Avian influenza (commonly referred to as avian flu or bird flu) is a disease that spreads through bird populations and can cause significant death of those birds, as reported by the RSPB in the United Kingdom. The disease tends to be classed into two groups, referred to as low pathogenic avian influenza (LPAI) and highly pathogenic avian influenza (HPAI). If birds are infected with a HPAI strain the animals may show a variety of symptoms, as listed in Table 7.3.

The active agent of avian influenza is caused by infection by a subtype of the influenza A virus (IAV).[50] A history of avian flu was written by Lycett et al. in 2019.[51] They start by reminding us of the human pandemic of 1918, which caused the death of about 50 million people. It was caused by the H1N1 strain of the virus. Like the bird flu virus, which is the H5N1 strain, the virus is a negative sense single-stranded RNA virus. The genome is segmented into eight sections, which have a size of 890 to 2,341 bases (or nucleotides). These viruses have a high mutation rate, meaning their infectivity can change relatively quickly. The disease has been known for a long time, being recorded back in the 1800s, but it was referred to as fowl plague. In 1955 Schafer[52] published a paper showing that the disease was caused by an influenza A virus. Lycett et al. go on to list several bird flu outbreaks, often caused by an H7 strain of the virus. It appears that water-associated birds can act as a natural reservoir for these viruses, and therefore they are often found in gulls, terns, duck, and geese, for example. However, migratory birds may carry the virus vast distances. In July 2023, rangers at Long Nanny in the United Kingdom were said to be "heartbroken"[53] when hundreds of Arctic terns were found dead with suspected bird flu, highlighting how migrants may be infected. However, the virus may cause an asymptomatic infection, so it is not always clear if a bird is carrying the virus.

The spread and impact of avian influenza is a serious problem. The present bird epidemic has been compared to the deaths in the 1950s and 1960s caused by dichlorodiphenyltrichloroethane (DDT), which is an organochloride for use as a pesticide. It was first made by Othmar

Table 7.3 Possible symptoms of avian influenza, as listed by the UK Government website[49]

Possible symptoms to be aware of
Lethargy and depression; lying down and unresponsiveness
Lack of coordination
Runny eyes, or eyes closed
Swollen head
Eating less than usual; sudden increase or decrease in water consumption
Head and body shaking
Drooping of the wings; dragging of legs
Twisting of the head and neck
Swelling and blue discolouration of comb and wattles
Haemorrhages and redness on shanks of the legs and under the skin of the neck
Breathing difficulties
Discoloured or loose watery droppings
Fever; increase in body temperature
Stop or significant drop in egg production
Sudden death

Zeidler, an Austrian chemist, in 1874, but it became commonly used in the middle of the 20th century. Despite its intended use it caused the mass death of many other species, including birds. The shells of the birds' eggs were produced too thin if the birds were poisoned by DTT, and many eggs were simply crushed in the nest. An influential book was written on the topic by Rachel Carson (1907–1964)[54] in 1962 called *Silent Spring*.[55] The present bird flu outbreak has been described as a new silent spring,[56] emphasising the perceived scale of the problem.

In December 2022, it was stated that the avian influenza epidemic was "The equivalent to our Covid pandemic,"[57] and that it was not going to stop in the near future. It was a problem that would continue through 2023 at least. There is likely to be a significant loss of birds in the wild. But birds, such as chickens and turkeys, are a major global industry. For Christmas 2022 there were serious concerns that the supply of turkeys would not keep up with demand as bird flu affected breeding farms.[58,59]

On 17 October 2022 the whole of the United Kingdom was declared to be an Avian Influenza Prevention Zone (AIPZ),[60] and those who were keeping birds had to follow strict guidelines. In England and Wales birds had to be housed to protect them from bird flu[61] and, importantly, prevent captive animals interacting with wild birds so that the flu virus did not spread.

Avian influenza is going to be around for a while longer. And the worry is about it infecting humans, which does happen. As of 1 October 2022, the British government was reporting that there were 151 confirmed cases of (HPAI) H5N1.[62] 134 of these were in England, but all of the UK nations were affected. In December 2020, a UK newspaper reported: "Since 2013 there have been 1,568 human cases of avian influenza and 616 deaths worldwide from the H7N9 strain."[63] The worry is that this will significantly increase as the bird epidemic continues in 2023.

A spokesperson for the Hawk Conservancy Trust, UK,[64] where there is a large bird population, said, "We are maintaining our biosecurity protocols and in fact are expanding these measures with forward planning to keep them in place" (Figure 7.2).

Therefore, avian influenza is a disaster for bird populations, whether they are in captivity or in the wild, but is it also a worry for human health? Alarmingly, if the statistics above are representative of the danger of this disease to humans, this would mean a mortality rate of 39%, more than 100 times that of COVID-19 and close to that of Ebola. As the epidemic in the bird populations continues, care will

Figure 7.2 Photograph of a Verreaux's Eagle Owl (or Milky Eagle Owl, sometimes called a Giant Eagle Owl: *Bubo verreauxii*); it is the largest African owl. Original photograph by J.T.H., but made black and white. Picture was captured at the Hawk Conservancy Trust, UK.

have to be taken by all those working in and around birds, whether they are in the farming industry, whether they are working in bird conservation, or whether it is a family keeping a few chickens at home for eggs, and possibly meat. In 2017 Taubenberger and Morens[65] said: "How IAVs evolve, switch hosts, and stably adapt to new hosts remain poorly understood." This does not sound too optimistic for containing a bird flu outbreak.

Interestingly, concerns around the re-establishment of mink farms, so devastated by the COVID-19 pandemic, have raised issues about the spread of other diseases, including avian influenza.[66] The worry is that the virus will spread in mammals rather than birds. It is stated that "H5N1 appears to have spread through a densely packed mammalian population and gained at least one mutation that favours mammal-to-mammal spread." Such reports highlight the danger of farming animals in such tight conditions. Conditions that encourage the spread and creation of variants of viruses may be good for more than one virus type, such as SARS-CoV-2 and H5N1. Clearly this situation needs to be carefully monitored in the future, else there is the potential for creating a new epidemic in the human population.

The Atlantic headlined an article with "We have a Mink Problem," pointing out that:

> Foxes, bobcats, and pigs have fallen ill. Grizzly bears have gone blind. Sea creatures, including seals and sea lions, have died in great numbers.[67]

They go on to say that the biggest problem is mink, and in Spain in October (one assumes 2022 as the article was February 2023) thousands of farmed minks were killed by bird flu and thousands more were then euthanised. It was thought that the virus had mutated and that this was the first major mammal-to-mammal outbreak. Echoes of COVID-19 could be seen here, where minks were euthanised in great numbers as the SARS-CoV-2 virus spread.

Furthermore, it has been reported that avian influenza has spread to otters and foxes in the United Kingdom,[68] and such animals are widespread across the nation. Others report infection in sea lions,[69] with a particularly large outbreak in Peru where at least 700 sea lions have been found to have died because of avian influenza infection.[70] It is thought that transmission to such animals is through the eating of infected birds. Clearly this virus can be spread relatively easily and can infect a wide range of mammals, and, therefore, there needs to be vigilance moving forward.

Certainly, headlines such as from the *Daily Mail* in February 2023 make this seem like a sensible thing to do:

"Bird flu 'may mutate to kill more than 50% of humans.'"[71] Sadly, as reported in February 2023, the first case in Cambodia for nine years claimed the life of an 11-year-old girl.[72] However, sequence analysis of the virus from the girl revealed that it was from the clade 2.3.2.1c, which had been in Cambodia since at least 2013.[73] It was not the same strain, i.e., 2.3.4.4b, which is causing such havoc in the bird population at present.

In April 2023, the first death from avian influenza in the country was reported in China.[74] A 56-year-old woman, from Guangdong Province in southeastern China, was reported to have died from H3N8 bird flu, whilst two other people were also said to be infected. The woman had several underlying health issues, and it was thought that the virus would not spread easily from human to human.

With bird flu clearly having an effect on humans in more than one country, it seems that biosecurity and vigilance here are key, and swift action, including virus genome sequencing, is needed if infections spread across the human population.

7.4.2 Mpox, is this a worry?

Bird flu is a serious worry, but it was not the only infectious disease that was in the media in early 2023. A second story centres around monkeypox, now referred to as mpox.

An overview of mpox was written in 2022 by Moore et al.[75] In an extract from this paper, it says that it was:

> first isolated and identified in 1959 when monkeys shipped from Singapore to a Denmark research facility fell ill. However, the first confirmed human case was in 1970 when the virus was isolated from a child in the Democratic Republic of Congo suspected to have smallpox.

The virus that causes mpox is related to that which leads to smallpox. It is a double-stranded DNA virus that exists in two types, or clades, i.e., Clade I and Clade II. Clade I viruses are more likely to lead to fatalities.[76]

Mpox is referred to as a self-limiting illness, although the R_0 number may be as high as 2.13.[77] Others think that the R_0 is less than 1, meaning that it would naturally dwindle out instead of raging through a population. Most people recover, although more severe symptoms can become manifest, and some people have died from mpox. The main symptoms are listed in Table 7.4 (Figure 7.3).

In June 2022 four possible scenarios were suggested of how the disease might play out in the future.[79] In the best case it may last a few months, perhaps dying out after about 3–4 months, with daily cases peaking at less than 30, and then virtually disappearing. In the second scenario it may

Table 7.4 The main symptoms of mpox, as listed on the UK Government website[78]

Symptoms of mpox	
Initial symptoms may include:	Fever
	Headache
	Muscle aches
	Backache
	Swollen lymph nodes
	Chills
	Exhaustion
	Joint pain
1–5 days after fever	Rash, starting on face but spreading across body

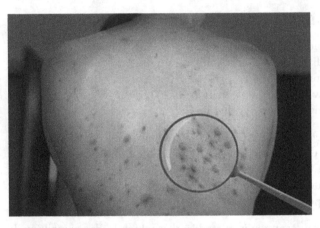

Figure 7.3 A patient with mpox, showing lesions on the skin. Photograph from Shutterstock, courtesy of Pavlova Yuliia, but made black and white.

last nearly a year, peaking at about 60,000 daily cases, but then die away. In the third case it peaks briefly at about 30 cases per day and then levels out at about ten cases a day, but never goes away. In the fourth scenario there is a small outbreak, peaking at 30 cases a day, then dying away, but then coming back as repeated epidemics at some future point, which is not specified. All this depends on how well the virus can spread between humans, between animals, and between different species.

Even though there appears to be some doubt about how the mpox will develop, or not, the WHO considered in February 2023 declaring the end of the mpox emergency.[80] The worry was that if this happened the Western and richer states would lose interest, thinking that their populations were safe. Research, monitoring, and funding would decrease, leaving some countries, particularly in Africa, vulnerable to the epidemic.

There are two main issues here. Although the main symptoms listed in Table 7.2 may not seem too bad, many people will die of mpox. The fatality rate can be as high as 11% across human populations, but for young people it can be higher. The average death rate is normally 3% to 6%.[81] In February 2023, it was reported that there were 85,922 cases worldwide in the latest outbreak,[82] and if we assume a fatality rate of 3% that would equate to 2,577 deaths. The good news is that the present outbreak is due to a Clade IIb virus, and this is much less fatal than a virus from Clade I. Therefore, such numbers of deaths are unlikely. However, if people do not die there can be severe outcomes, including sepsis and eye infections, that can lead to loss of vision.

Second, there is no reason to assume that the virus will not mutate. And as long as it is circulating the chances of mutation increase. Humans and animal populations make good reservoirs for viruses, and if a more virulent strain emerges the transmission of the virus through humans or animals may be greater than it presently is.

Having discussed the virus with an anthropocentric lens, what about the animals? Mpox can persist in rodent populations (squirrels, rats, and mice), where it seems to be mainly asymptomatic. Non-human primates can get the disease with human-like symptoms. Other animals known to have become infected include anteaters, shrews, and hedgehogs. It is thought that infection of birds, reptiles, and amphibians is very unlikely.[83]

Will mpox cause the next pandemic? It seems unlikely to be a major problem, but it certainly should not be dismissed. It will circulate in both human and animal populations and may re-emerge in new human communities. Along with bird flu, mpox monitoring and research needs to be continued to keep both us and our ecosystems safe.

7.5 How to mitigate effects on conservation and use of PPE

A vast quantity of PPE was used worldwide during the COVID-19 pandemic. Of course, this is necessary to keep humans safe, but as discussed previously, this does have an effect on animals and the wider ecosystems.

Some of the obvious solutions to PPE use, but which are not as easy to implement, may be:

- Use less PPE. However, people need to be protected and the natural thing to do is to use it whether needed or not, as being protected is better than being infected. There must have been a considerable amount used that did not need to be, but stopping people from trying to protect themselves is not a sensible approach. However, there may be situations for which PPE may not be needed in the next pandemic, and automatically reaching for gloves and masks may not be necessary.

■ Use less plastic in PPE. However, in many cases this would need a re-think of the materials used. Plastic is easy to produce and to mould and it is relatively cheap. It is also light and easy to transport. However, there are alternatives available, such as promoted by *A Plastic Planet*,[84] and research and development in this area needs to make sure that non-plastic PPE is not only able to be produced in large enough quantities but is also fit for purpose.

■ Make PPE disposal in a sustainable manner, or make it reusable. Either way, the quantity taken to landfill sites would be reduced. If the PPE were able to break down naturally in the environment there would be less impact on animals, and if it was reusable there may be less littering to start with.

In their paper, De-la-Torre et al.[85] complain about the "lack of environmental awareness" when it comes to plastic pollution, and they go on to say:

Extended and long-lasting marine litter monitoring is required to set ground for the development of policy briefs aimed for a better PPE waste management in Peru.

Although this paper is focused on Lima in Peru, this makes sense, and such sentiments and ideas can be translated across the globe. They also go on to propose ways that PPE can be considered in the future and say:

Additional alternatives recently investigated involve the recovery of liquid fuels from PPE waste through pyrolysis. This may be promising recycling route for PPE after recovery. Redesign of conventional PPE may include natural biodegradable materials, such as PLA and starch, to facilitate solid waste management. Also, promoting reusable masks.

These are all sensible suggestions and should be considered for the future development and use of PPE. However, this needs forward planning. In the COVID-19 pandemic, at least in the United Kingdom, there was a scramble to obtain enough PPE,[86] and many contracts given by the government have since been questioned on ethical grounds. Such actions are not conducive to obtaining biodegradable and sustainable PPE, and they need to be avoided in the future.

7.6 Ethics considerations when considering SARS-CoV-2 and animals

As we have argued several times already, the response to COVID-19 was extremely human-centric, and perhaps to a large extent that was the correct approach, but as we will argue later, in the future we need a One World/One Health approach to disease management.

The COVID-19 pandemic had a huge testing programme – track and trace – but this did not, in general, include animals. Should more have been done for the animals, either companion individuals or in the wild? This would have taken away critical resources, both time and materials, from testing and tracking humans. Some countries were struggling to manage the pandemic as it was, and they did not have the finances or the resources to worry about the local animals. However, we live together, so perhaps in some ways it could be argued that this was short-sighted.

On a similar note, the human population was vaccinated, at least in the wealthy countries. Some animals were, the vast majority were not. Should more have been done to protect the animals? Again, this would have taken valuable resources away from the human treatment regimes, and it can be asked: Would that have been the right thing to do? Should we vaccinate a gorilla population and not the humans who may be close by? And which animals do we target? It is probably not a coincidence that the World Wildlife Fund (WWF) uses a giant panda as its logo.[87] It is easy to care about the cuddly animals. But what of the ugly and dangerous ones? Would you put the life of a Komodo dragon high on your list to be saved, considering how dangerous they are? Therefore, which animals are worth saving and which are not? Surely, as we share the same rock spinning around the sun, are we not all in this together and are not all animals equal?

7.7 One World/One Health/One Medicine

As we move into the future, we should be considering a One World/One Health/One Medicine approach to how we manage the potential and reality of new pandemics.[88,89,90] As pointed out in Chapter 1, COVID-19 had been met with a very anthropocentric view. This blinkered approach is not helpful as we move forward.

The COVID-19 pandemic seemed to come as a surprise to many, even though many years before such an incident was predicted. As discussed, the pandemic was an event that had animals as an integral part. Sujit Sivasundaram,[91] taking a historical approach, says that we should "begin with the premise that it is not a bolt from the blue nor an unexpected disaster." He goes on to say: "It is necessary to decentre the anxious human subject" and that "zoonotic transfer occurs where relations between humans and animals have been unstable or where they are entering a new phase of contact." Such discussion suggests that an anthropocentric attitude is not helpful when considering events such as pandemics and that keeping animals in the discussion is very important. However, this is not always achieved. As Irus Braverman has said:

> Despite the expansiveness of the term, however, scholars have pointed out that much of the research on One Health is still too anthropocentric — and, relatedly, also too vertebrate-centric and eukaryo-centric — and that it has typically focused, accordingly, on zoonotic and vector-borne diseases and on health services that integrate only human and animal health, without considering their environment. Indeed, although One Health prides itself on aspiring to decenter the human, its practices have not always followed suit.[92]

She goes on to refer to this being apparent "in the context of industrial animal agriculture, which is responsible for the annual killing of tens of billions of terrestrial animals and a trillion aquatic animals" and "the mass killings of nonhuman animals as means of disease prevention."

This bears closer examination in the context of current policy debates (Figure 7.4).

One Health has had many definitions over time (a useful timeline is presented by the CDC[94]), starting with One Medicine (which can at its simplest be conceptualised as strong, two-way links between veterinary and human medicine professionals), but it is much broader than that, for example in relation to what those links can actually achieve, such as two-way reciprocal benefit and building a focus on studies of naturally occurring disease in animals rather than a focus on laboratory animal models. This is a necessary but not sufficient underpinning for a One Health approach that is not anthropocentric, and where the benefits can be felt by animals as well as humans. The One Health High-Level Expert Panel (OHHLEP) has set out definitions[95] and principles of One Health, including, within the definition:

- One Health is an integrated, unifying approach that aims to sustainably balance and optimize the health of people, animals, and ecosystems.

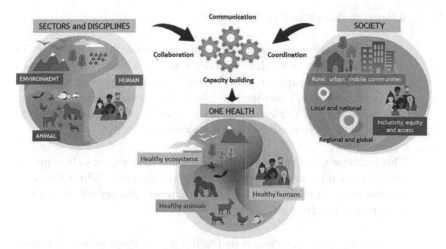

Figure 7.4 An infographic summing up the interconnection between people, the environment, and animals and keeping all aspects healthy. Image developed by the One Health High Level Expert Panel.[93]

- It recognizes that the health of humans, domestic and wild animals, plants, and the wider environment (including ecosystems) are closely linked and interdependent.

and, within the principles:

3. socioecological equilibrium that seeks a harmonious balance between human–animal–environment interaction and acknowledging the importance of biodiversity, access to sufficient natural space and resources, and the intrinsic value of all living things within the ecosystem; 4. stewardship and the responsibility of humans to change behaviour and adopt sustainable solutions that recognize the importance of animal welfare and the integrity of the whole ecosystem, thus securing the well-being of current and future generations.

This appears to recognise the joint importance of humans, animals, and ecosystems. However, as we will see below, this does not seem to have permeated fully into the current global pandemic preparedness agenda.

OHHLEP has produced a White Paper/Opinion piece[96] about prevention of zoonotic spillover, which they define here as:

Prevention of spillover in the context of this paper refers to preventing the critical first step, i.e. preventing a pathogen from transferring from animals to humans.

Related to that definition is a footnote:

> While this paper specifically addresses pandemic prevention in humans, in line with the OHHLEP One Health definition endorsed by the Quadripartite it is important to note that pathogen spillover from humans to other species or between other species facilitated by human activity (e.g., wildlife trade) can also have devastating impacts on wild and domestic animal populations.

This footnote is fundamentally true, but it does not appear to be a proper One Health approach to relegate animals to a footnote. Whilst the justification may rest in the fact that by far the most spillover is from animals to humans, they later go on to say:

> An important principle is that spillover of pathogens from a natural source only occurs at risky exposure interfaces between humans, animals and the environment, such as direct or indirect contact between the pathogen (e.g. via an infected host/environment) and people. Animals and biodiversity do not present an inherent risk per se; risk is created by human behavior that places humans and other species in risky contact that increase chances for spillover.

Again, this is patently correct, and, in that context, it would surely make sense to consider a definition of spillover that encompasses the spillover of animal-to-animal transmission, such as avian influenza into mammals or from humans to animals; or such as COVID 19 from humans to minks, gorillas, and tigers; or in future, potentially a virus passing from humans to bats, and there mutating into something that threatens humans and other animals.

There is also a later statement that could equally be applied to avoiding spillover from one animal species to another, especially in the context of anthropogenic influence:

> Specific factors related to hunting, capturing, farming and slaughter/preparation of wild animals; intensive/high density livestock farming especially linked to inadequate biosecurity; trade in live animals and animal products; deforestation, extractive industries, and encroachment into wildlife habitat; agricultural expansion and intensification; urbanisation and habitat fragmentation are often important in shaping risk. Overarching drivers, such as climate change, food security, basic animal and human health, and animal welfare practices, poverty, and socioeconomic inequalities, should also be considered in the prevention of spillover.

As a response to the pandemic, it is now (at the time of writing) planned that there should be an "international instrument" under the auspices of WHO with the aim of strengthening "pandemic prevention, preparedness and response." There is currently a "zero draft" of this document, which will form the basis for negotiations by Member States on a "pandemic accord." This defines One Health in the following manner:

> One Health – Multisectoral and transdisciplinary actions should recognize the interconnection between people, animals, plants and their shared environment, for which a coherent, integrated and unifying approach should be strengthened and applied with an aim to sustainably balance and optimize the health of people, animals and ecosystems, including through, but not limited to, attention to the prevention of epidemics due to pathogens resistant to antimicrobial agents and zoonotic diseases.[97]

This appears on the face of it to include a concern for the well-being of animals, but later in the document the following appears under "One Health":

> The Parties, with an aim of safeguarding human health and detecting and preventing health threats, shall promote and enhance synergies between multisectoral and transdisciplinary collaboration at the national level and cooperation at the international level, in order to identify, conduct risk assessment of and share pathogens with pandemic potential at the interface between human, animal and environment ecosystems, while recognizing their interdependence.

If the aim is still solely to be about safeguarding human health, it does appear that the point of us living in an integrated ecosystem (as reflected in the definitions and principles espoused by OHHLEP, as cited above) may not be as embedded as might be imagined. The outcomes are still to be about human health; the bottom line here appears to be that animal health only matters insofar as it affects human health.

One might counter that, of course, we need to prevent the horrendous suffering of another pandemic for humans, that animals are not the target here for the WHO, people are. However, surely the protection of animal health and well-being is an important target for this very same protection of human health? If we do not keep minks in such close quarters, does it not make widespread disease amongst minks, where a virus could mutate and pass back to humans, less likely? If we did not trap wildlife, keep it in horrendous conditions, then eat it, would that not give rise to less risk for humans? If we manage the flow of effluent into our rivers significantly better, will it not only benefit the ecosystem and the wild creatures that exist in it, but will it impact upon the infectious agents present in our human environment? And of course, all of these sorts of things will be considered, that is taken as read (because they have a potential impact on human health). This is not at all a questioning the competence of those engaged in this crucial endeavour – it is just seeking to broaden out the viewpoint a little.

The point being raised here is that the ultimate outcome should not *only* be for human health. Do we not have a moral obligation, where animals are being made sick by human action, whether that be ecosystem degradation due to anthropogenic climate change and pollution, habitat destruction, profit driven, or low welfare, farming practices, to at least consider animal health and well-being as part of our One Health pandemic "cunning plan"?[98] Would we not consider, for example, it to be important that gorillas or Scottish wildcats not be made extinct by a pandemic? Or that if we have an avian influenza pandemic, we may end up culling vast numbers of birds (because if human health was at stake, that seems likely), some of which species may thereby be made extinct? Not caring about the well-being of animals, for its own sake, could be seen to have got us into this mess in the first place. If we (all, globally) had really cared about the well-being of animals, they would not be held in the kind of conditions prevalent at "wet markets" where animals are sold. If we did not factory farm animals in inhumane conditions, crowded together as they live and die, well, if we cared about that, perhaps we may avoid the next pandemic.

It is not only an ethical and moral imperative but also a crucial step in pandemic prevention to consider the health and well-being of animals as important in and of themselves, not just where they might pose a risk to humans. This is not least because we will not always be able to judge when that risk will actually explode, as we have seen with the COVID-19 pandemic – no one knew this disease was about to be released on the world, but perhaps if we had had a more animal welfare–centric approach, that would have been a crucial preventative factor. It is just as crucial, of course, to consider how both humans and non-human animals are interacting within ecosystems that are, increasingly, being themselves affected by negative anthropogenic change. That one may be the hardest nut to crack.

This view of the current "Zero draft" wording appears to be shared, at least in part, by others. For example, the minutes of the eighth meeting of the OHHLEP[99] includes, "The pandemic instrument focuses on human pandemics and is therefore anthropocentric, how OHHLEP could influence the One Health definition and the value it could add was addressed." Let us hope they are successful and the future is better than we might anticipate.

7.8 Plants and ecosystems

The discussion in this book has concentrated on the effect of COVID-19 on animals, and as SARS-COV-2 is a virus that infects animals, probably exclusively vertebrates, this is understandable. Many of the indirect effects are seen on animals too, such as PPE entanglement, so it seems appropriate that such discussion should take centre stage. However, animals, including we humans, do not live in isolation; rather, we live as part of a complex ecosystem and the effects of an epidemic/pandemic will radiate across such ecosystems.

Plants are an integral part of any ecosystem, so what were the effects of the pandemic on these organisms?

In simplistic terms as animals are affected plants will be too. As discussed before, the roaming range of many animals was increased and therefore the foraging of such animals would have also been more wide ranging. On the other hand, the roaming of many humans would have been curtailed – at least for short periods of lockdown – and therefore no doubt foot fall damage to plants would have decreased in many places. On the other hand, increased litter, such as PPE, would have been damaging to many plants, as they are screened from light by fallen masks and gloves. Widespread spraying of disinfectants would have been detrimental to land plants, and runoff into rivers and streams, and eventually to the sea, would have been harmful to aquatic plants.

However, as with all the discussions above, some perspective is needed here. Plants are quite resilient to damage, especially species such as grasses, and it would not be long before what would have no doubt been a minor detriment to plants would have been restored.

Having said that, the pandemic did have some impact on the agriculture industry. Supply chains, and staff available for fertiliser and pesticide use did decrease, with an ensuing reduction in crop yields and increase in crop damage. There were some issues also raised about biosecurity of plants and whether this may have a longer-term impact. For a more in-depth discussion on these issues, an editorial was written by Lamichhane and Reay-Jones,[100] and this points to a series of papers on the topic.

Indoor plants, especially ones left in offices and industrial settings, were also sometimes found to be negatively affected.[101] Some had unusual water regimes, either too much or too little, as offices were left abandoned. Light levels were affected as auto switch-off systems were not triggered in empty corridors, so plants were left in the dark. Extra cleaning often saw overspill into containers in which plants were growing. No doubt many people returned to offices with their plants either very unhealthy or dead.

On the positive side of the discussion of plants and the pandemic, it has been suggested that plants may be useful for vaccine production, especially alfalfa (*Medicago sativa*).[102] Plants may also be useful as COVID-19 medicines.[103] Therefore, integrating plants into our thoughts about a human viral disease is important.

Plants, therefore, were not immune to the pandemic, even if the virus was one that infected animals. However, in most cases, either in the wild or in domestic/commercial settings, plants are easy to propagate and grow, and therefore there is less long-term impact from a pandemic, and, as plants are not sentient, there is much less emotional attachment. Most people would have ignored the fate of the plants around them during the pandemic, and unless directly affected by the lack of food, or some other unanticipated consequence such as wildfires that cannot be controlled because of human or animal health issues, this will no doubt be the situation in any future pandemic too. However, plants are central to our food chains and ecosystems, so considering any detrimental effects is important.

7.9 Conclusions

As WHO has announced that the second phase of the investigations into the origin of SARS-COV-2 is now cancelled,[104] it is now unlikely that the true source of this virus will ever be known. It may have originated in a bat in a cave in southwestern China, but it may have been engineered in a laboratory before being released,[105] almost certainly accidentally if that was the route. However, we may never know, which will not be useful for planning mitigating actions for any future similar viral spread. Afterall, there will be a next one, be it SARS-related, bird flu, mpox, or something as yet unknown. Societies around the world need to be prepared. We need to know where such viruses are emanating from and then what to do about it when it happens. As part of that planning we need to consider the wider environment around us, and that includes the animals, especially if animals are the origin source of the virus. If it was confirmed that SARS-CoV-2 had originated from a bat we should be able to determine how human/bat interactions need to be managed in the future. Without such knowledge, and without putting such knowledge into action, another virus from another bat may be responsible for the next pandemic.

There has been a vast amount of scientific literature on the COVID-19 pandemic. A crude measure can be gained by putting the term "COVID-19" into a Google Scholar search (with plus citations turned off), and the algorithms return an amazing 4,950,000 results (as of December 2022). Doing the same search with "COVID-19" and "Animals'" still gives

768,000 papers, and there is no way any text can encompass this mass of work. Therefore, this text gives somewhat of a snapshot of the situation as regards SARS-CoV-2 infections and animals.

Even with the caveat there are a number of conclusions that can be reached, which can be summarised in a series of bullet points:

- COVID-19 caused millions of human deaths, estimated as of December 2022 as being 6,678,922.[106]

- The virus causing COVID-19 was a coronavirus that scientists named SARS-CoV-2.

- SARS-CoV-2 almost certainly originated in a bat in China. The first case of COVID-19 was in Wuhan, but how the virus "jumped" from an animal to a human may never be established for sure.

- Many animals were infected, and several died as a direct consequence of being infected with SARS-CoV-2. However, the numbers recorded were much lower than that for humans.

- In most animals the symptoms following SARS-CoV-2 infection were mild and the animals recovered relatively quickly.

- The presence of species-specific coronaviruses (e.g., canine and feline) makes it difficult to know if a coughing animal is suffering from a SARS-CoV-2 infection or another virus. Even with humans, if someone coughs on a train, others in the carriage may automatically think it is COVID-19, but other pathogens are still out there, such as cold and influenza.

- The total infection rates and deaths from direct infection in animals will never be known, and they are likely to be significantly higher than reported.

- Molecular biology and bioinformatics have failed to produce a robust method for predicting which animals are susceptible to SARS-CoV-2.

- The infections of animal species in the real world did not reflect what was expected from the molecular studies. Why were mustelids so affected? Why deer and not other ungulates?

- In many respects the pandemic could have been much worse. Obviously, a contentious thing to say when nearly seven million humans died. The human death rate could have been much higher, and most of the animals that we rely on for food security were not affected. Cows, pigs, sheep, chickens, etc. were not a major issue. Their farming, slaughter, and transport to markets may have been problematic, but, in most cases, food supplies continued to a sufficient level.

■ Vaccines against COVID-19 were relatively quickly developed and saved millions of lives. They were used on some animals too.

■ Some of the mitigating actions to protect humans were harmful to animals. This included the use of detergents and PPE (as well as culling, such as in the case of minks).

The future needs to be considered here too. There will be more epidemics or pandemics. Some of these may be caused by SARS-like viruses or other coronaviruses. Other diseases will be caused by unrelated viruses or bacteria. Whatever the cause there are a series of considerations that may make our world a safer and fairer place:

■ We need to understand better how viruses "jump" from animals to humans within the various contexts for animal-human interaction. In future, there needs to be more transparency in emergent outbreak information than we saw in the early stages of this pandemic such that disease outbreaks can be more effectively controlled. Wet markets of the type found in Wuhan, which sell live animals, ought at the very least to be better regulated, and that regulation enforced.

■ Governments around the world need to be better prepared. It will happen again; it is not a case of if, but when.

■ A better and more fair distribution of materials is needed. This includes vaccines and PPE. Why should rich countries be able to protect their populations whilst other areas of the world struggle?

■ A One World/One Health policy is the clearly sensible way forward. However, just like trying to face climate change, there are major changes that are needed, both in ethos and in funding. However, as already discussed, we share our planet with all the other animals (and plants) and we have nowhere else to go. We wreck our one rock, and we are in trouble.

■ Animals are important to us. They are part of our ecosystem. They provide food for millions of people. They give us joy as companion animals. We love to see them in the wild and in conservation societies and zoos. We need to develop an ethos, globally, in which animal welfare matters. Even for those who would not consider this a priority for its own sake, there is a compelling argument that poor animal welfare practices could contribute to the emergence of the next human pandemic.

It is encouraging that in April 2023, the United Kingdom announced the Respiratory Virus and Microbiome Initiative. The stated aim is:

> to establish capabilities for large-scale genomic surveillance of respiratory viruses such as influenza virus and respiratory syncytial virus (RSV), and to survey for as yet unknown pathogens.[107]

The initiative has a specific aim to "inform early warning systems for new pandemic threats." Hopefully, such work will help to predict and then control any future pandemics, whether they are caused by a SARS-like virus or some other pathogen. Also, as already mentioned, the United Kingdom announced a "genomics transformation,"[108] so it is clear that nations are looking for a technological solution to the rise of the next pandemic.

It may sound semantic, but there has been an argument that the pandemic should not be called that at all. Rather, it should be termed "panzootic."[109] Agnelli and Capua point out that as of October 2022, around the world there have been 675 natural outbreaks of SARS-CoV-2 infection in animals, and a term such as "panzootic" would be more representative of the situation, giving less of an anthropocentric impression of how the virus had its effects. Perhaps adoption of such a term may keep animals in mind as humans tackle the next pandemic. After all, there are already concerns about mpox and avian influenza, as discussed previously, and both such diseases clearly involve animals. The next pandemic will no doubt originate in an animal, and it will affect humans and animals as it spreads across geographies and species. The next outbreak is likely to be panzootic.

This can be illustrated in relation to the Pandemic Influenza Preparedness Framework,[110] in which the Partnership Contribution High-Level Implementation Plan III 2024–2030 says the following:

> the increasing impact of environmental degradation and climate change on human health has shown a key need to integrate animal and environmental health into how surveillance systems assess risk. As noted in the Joint Risk Assessment Operational Tool of the Tripartite Zoonoses Guide: bringing together national information and expertise from all relevant sectors for the joint assessment of health risks from zoonotic disease allows all sectors, acting together, to evaluate fully, understand and manage shared risks at the human–animal–environment interface. This perspective recognizes the ways in which environment and animal health can spill over into humans through zoonoses, thereby increasing the risk of the emergence of influenza viruses of pandemic potential.

It is possible that, if one regarded this as a panzootic, there would be more emphasis placed on surveillance and identification of spillover between other animal species, not just humans.

Notes

1 Ng, M.K. (2020) Sustainable development goals (SDGs) and pandemic planning. *Planning Theory & Practice, 21*, 507–512.
2 Nature: Did flu come from fish? Genetics points to influenza's aquatic origin: https://www.nature.com/articles/d41586-023-00558-4?utm_source

=Nature+Briefing&utm_campaign=22d7dd2757-briefing-dy-20230301&utm_medium=email&utm_term=0_c9dfd39373-22d7dd2757-45565222 (Accessed 07/03/23).

3 CNN: Biden declares the pandemic over. People are acting like it too: https://edition.cnn.com/2022/09/19/politics/biden -covid-pandemic-over-what-matters/index.html (Accessed 23/12/22).

4 Along with Dr Maria Elena Bottazzi, Dr Peter Hotez was nominated in 2022 for the Nobel Peace Prize. ABC13: 2 Houston scientists nominated for Nobel Peace Prize for work in COVID vaccine development: https://abc13.com/nobel-peace -prize-nominees-2022-price-nominations-houston-researchers-nominated-for /11530065/ (Accessed 06/01/23).

5 Guardian: Nobel-nominated vaccine expert warns of Covid complacency: 'We're still losing too many lives': https://www.theguardian.com/world/2022/dec/23/covid-cases-death-rate-risk-peter-hotez (Accessed 06/01/23).

6 Guardian: Chinese city seeing half a million Covid cases a day – local health chief: https://www.theguardian.com/world/2022/dec/24/chinese-city-seeing-half-a -million-covid-cases-a-day-local-health-chief (Accessed 24/12/22).

7 US News: Report: China's COVID-19 infections surge, reach 37 million in single day: https://www.usnews.com/news/health-news/articles/2022-12-23/report -chinas-covid-19-infections-surge-reach-37-million-in-single-day (Accessed 24/12/22).

8 Bloomberg UK: China estimates Covid surge is infecting 37 million people a day: https://www.bloomberg.com/news/articles/2022-12-23/china-estimates -covid-surge-is-infecting-37-million-people-a-day?leadSource=uverify%20wall (Accessed 24/12/22).

9 BBC: What is known about new Covid variant XBB.1.5? https://www.bbc.co .uk/news/health-64164306 (Accessed 06/01/23).

10 Guardian: Plans for UK 'genomics transformation' aim to act on lessons of Covid: https://www.theguardian.com/society/2023/may/16/plans-for-uk-genomics -transformation-aim-to-act-on-lessons-of-covid (Accessed 16/05/23).

11 UK Government: Long-term effects of coronavirus (long COVID): https://www .nhs.uk/conditions/coronavirus-covid-19/long-term-effects-of-coronavirus -long-covid/ (Accessed 24/12/22).

12 IFLScience: Long COVID's 200 symptoms have been narrowed down to just 12: https://www.iflscience.com/long-covids-200-symptoms-have-been-narrowed -down-to-just-12-69140 (Accessed 30/05/23).

13 CNBC: Long Covid medical costs average $9,500 in first six months, as patients become 'health-system wanderers': https://www.cnbc.com/2022/12/15/long -covid-medical-costs-average-9500-in-first-six-month-study.html (Accessed 23/12/22).

14 Live Science: Gorillas & humans closer than thought, genome sequencing reveals: https://www.livescience.com/18892-gorillas-humans-gene-sequence.html (Accessed 24/12/22).

15 Böszörményi, K.P., Stammes, M.A., Fagrouch, Z.C., Kiemenyi-Kayere, G., Niphuis, H., Mortier, D., van Driel, N., Nieuwenhuis, I., Vervenne, R.A., Haaksma, T. and Ouwerling, B. (2021) The post-acute phase of SARS-CoV-2 infection in two macaque species is associated with signs of ongoing virus replication and pathology in pulmonary and extrapulmonary tissues. *Viruses*, *13*, 1673.

16 Krouzecky, C., Aden, J., Hametner, K., Klaps, A., Kovacovsky, Z. and Stetina, B.U. (2022) Fantastic beasts and why it is necessary to understand our relationship – animal companionship under challenging circumstances using the example of long-Covid. *Animals, 12,* 1892.

17 BBC; Coronavirus: This is not the last pandemic: https://www.bbc.co.uk/news/science-environment-52775386 (Accessed 26/02/23).

18 Reperant, L.A. and Osterhaus, A.D. (2017) AIDS, Avian flu, SARS, MERS, Ebola, Zika… what next? *Vaccine, 35,* 4470–4474.

19 Nature: Five ways to prepare for the next pandemic: https://www.nature.com/articles/d41586-022-03362-8 (Accessed 26/0/23).

20 Furuse, Y. (2021) Genomic sequencing effort for SARS-CoV-2 by country during the pandemic. *International Journal of Infectious Diseases, 103,* 305–307.

21 BBC: Global Covid vaccine rollout a stain on our soul – Brown: https://www.bbc.co.uk/news/health-59761537 (Accessed 26/02/23).

22 US News: Students in Denmark go to school in Soccer Stadium: https://www.usnews.com/news/education-news/articles/2020-05-11/students-in-denmark-go-to-school-in-soccer-stadium-amid-coronavirus-pandemic (Accessed 07/03/23).

23 G7: A group representing the biggest seven economies in the world.

24 Lappan, S., Malaivijitnond, S., Radhakrishna, S., Riley, E.P. and Ruppert, N. (2020) The human–primate interface in the New Normal: Challenges and opportunities for primatologists in the COVID-19 era and beyond. *American Journal of Primatology, 82,* e23176.

25 Ross, A.G., Crowe, S.M. and Tyndall, M.W. (2015) Planning for the next global pandemic. *International Journal of Infectious Diseases, 38,* 89–94.

26 Cox, N.J., Tamblyn, S.E. and Tam, T. (2003) Influenza pandemic planning. *Vaccine, 21,* 1801–1803.

27 WHO: Global pandemic influenza action plan to increase vaccine supply: https://apps.who.int/iris/bitstream/handle/10665/69388/WHO_?sequence=1 (Accessed 26/02/23).

28 Kotalik, J. (2005) Preparing for an influenza pandemic: Ethical issues. *Bioethics, 19,* 422–431.

29 Smith, M.J. and Silva, D.S. (2015) Ethics for pandemics beyond influenza: Ebola, drug-resistant tuberculosis, and anticipating future ethical challenges in pandemic preparedness and response. *Monash Bioethics Review, 33,* 130–147.

30 Taubenberger, J.K. and Morens, D.M. (2010) Influenza: The once and future pandemic. *Public Health Reports, 125,* 15–26.

31 Thoradeniya, T. and Jayasinghe, S. (2021) COVID-19 and future pandemics: A global systems approach and relevance to SDGs. *Globalization and Health, 17,* 59.

32 UK COVID-19 inquiry: Resilience and preparedness: https://covid19.public-inquiry.uk/modules/resilience-and-preparedness/ (Accessed 07/07/23).

33 Evening Standard: Government defends decision to give Cheltenham Festival go-ahead in days before coronavirus lockdown: https://www.standard.co.uk/sport/horse-racing/government-cheltenham-festival-coronavirus-a4419376.html (Accessed 26/02/23).

34 UTMB: Tracking coronavirus in animals takes on new urgency: https://www.utmb.edu/one-health/news/news-stories/2022/05/23/tracking-coronavirus-in-animals-takes-on-new-urgency (Accessed 16/05/23).

35 Pratelli, A., Buonavoglia, A., Lanave, G., Tempesta, M., Camero, M., Martella, V. and Decaro, N. (2021) One world, one health, one virology of the mysterious labyrinth of coronaviruses: The canine coronavirus affair. *The Lancet Microbe, 2,* e646–e647.

36 Crary, A. and Gruen, L. (2022) *Animal Crisis. A New Critical Theory*. Polity Press, Cambridge. ISBN: 9781509549672.

37 Nature: Trove of new coronaviruses uncovered in bats — but threat is unclear: https://www.nature.com/articles/d41586-023-02151-1?utm_source=Nature +Briefing&utm_campaign=7e35f50b0b-briefing-dy-20230628&utm_medium =email&utm_term=0_c9dfd39373-7e35f50b0b-45565222 (Accessed 29/06/23).

38 Kane, S.A., Lee, Y.E., Essler, J.L., Mallikarjun, A., Preti, G., Verta, A., DeAngelo, A. and Otto, C.M. (2022) Canine discrimination of ovarian cancer through volatile organic compounds. *Talanta*, *250*, 123729.

39 Gao, C.Q., Wang, S.N., Wang, M.M., Li, J.J., Qiao, J.J., Huang, J.J., Zhang, X.X., Xiang, Y.Q., Xu, Q., Wang, J.L. and Liu, Z.H. (2022) Sensitivity of sniffer dogs for a diagnosis of Parkinson's disease: A diagnostic accuracy study. *Movement Disorders*, *37*, 1807–1816.

40 Rooney, N.J., Guest, C.M., Swanson, L.C. and Morant, S.V. (2019) How effective are trained dogs at alerting their owners to changes in blood glycaemic levels?: Variations in performance of glycaemia alert dogs. *PLoS One*, *14*, e0210092.

41 Maa, E., Arnold, J., Ninedorf, K. and Olsen, H. (2021) Canine detection of volatile organic compounds unique to human epileptic seizure. *Epilepsy & Behavior*, *115*, 107690.

42 Guardian: Winning by a nose: The dogs being trained to detect signs of Covid-19: https://www.theguardian.com/world/2020/jun/21/winning-by-a-nose-the -dogs-being-trained-to-detect-signs-of-covid-19 (Accessed 08/07/23).

43 Pirrone, F., Piotti, P., Galli, M., Gasparri, R., La Spina, A., Spaggiari, L. and Albertini, M. (2023) Sniffer dogs performance is stable over time in detecting COVID-19 positive samples and agrees with the rapid antigen test in the field. *Scientific Reports*, *13*, 3679.

44 Meller, S., Al Khatri, M.S.A., Alhammadi, H.K., Álvarez, G., Alvergnat, G., Alves, L.C., Callewaert, C., Caraguel, C.G., Carancci, P., Chaber, A.L. and Charalambous, M. (2022) Expert considerations and consensus for using dogs to detect human SARS-CoV-2-infections. *Frontiers in Medicine*, *9*, 3590.

45 Health Desk: How do symptoms vary by COVID-19 variants? https://health -desk.org/articles/how-do-symptoms-vary-by-covid-19-variants (Accessed 15/08/23).

46 Xiao, X., Newman, C., Buesching, C.D., Macdonald, D.W. and Zhou, Z.M. (2021) Animal sales from Wuhan wet markets immediately prior to the COVID-19 pandemic. *Scientific Reports*, *11*, 1–7.

47 BBC Sound: The Briefing Room – How worried should we be about avian flu? - BBC Sounds: https://www.bbc.co.uk/sounds/play/m001hx6s (Accessed 27/02/23).

48 RSPB: Avian Flu (Bird Flu): https://rspb.org.uk/birds-and-wildlife/advice/how -you-can-help-birds/disease-and-garden-wildlife/avian-influenza-updates/?from =morelikethis (Accessed 27/02/23).

49 UK Government: Bird flu (avian influenza): How to spot and report it in poultry or other captive birds: https://www.gov.uk/guidance/avian-influenza-bird-flu (Accessed 24/12/22).

50 There are four main types of influenza virus, termed A, B, C, and D.

51 Lycett, S.J., Duchatel, F. and Digard, P. (2019) A brief history of bird flu. *Philosophical Transactions of the Royal Society B*, *374*, 20180257.

52 Schafer W. (1955) Vergleichende sero-immunologische Untersuchungen über die Viren der Influenza und klassischen Geflügelpest [Comparative sero-immuno-logical studies on the viruses of influenza and classical avian influenza]. *Zeitschr. Naturforsch*, *10*, 81–91 (note, this paper is in German).

53   Guardian: Rangers 'heartbroken' after 600 dead Arctic tern chicks found in Northumberland: https://www.theguardian.com/environment/2023/jul/10/dead-arctic-tern-chicks-northumberland-long-nanny-bird-flu-suspected (Accessed 21/07/23).

54   Guardian: What the world can learn from Rachel Carson as we fight for our planet: https://www.theguardian.com/commentisfree/2021/oct/27/what-the-world-can-learn-from-rachel-carson-as-we-fight-for-our-planet (Accessed 24/12/22).

55   Carson, R. (1962) *Silent Spring*. Houghton Mifflin, Boston, MA.

56   Guardian: Deaths of thousands of wild birds from avian flu is 'new Silent Spring': https://www.theguardian.com/environment/2022/dec/23/deaths-thousands-wild-birds-avian-flu-new-silent-spring-aoe (Accessed 24/12/22).

57   Guardian: 'The equivalent to our Covid pandemic': Bird flu hasn't gone away and is still spreading: https://www.theguardian.com/environment/2022/dec/24/bird-flu-hasnt-gone-away-bleak-outlook-2023-aoe (Accessed 24/12/22).

58   BBC: Christmas turkey fears as England bird flu rules widened: https://www.bbc.co.uk/news/science-environment-63431588 (Accessed 24/12/22).

59   Anecdotally, sadly, and ironically, JTH was told by a member of supermarket staff that they would be throwing away turkeys on Christmas Eve as they had too many!

60   Scottish Government: Avian influenza (bird flu) outbreaks: https://www.gov.scot/publications/avian-influenza-outbreaks/ (Accessed 24/12/22).

61   UK Government: Bird flu (avian influenza): Housing your birds safely: https://www.gov.uk/guidance/bird-flu-avian-influenza-housing-your-birds-safely (Accessed 24/12/22).

62   UK Government: Bird flu (avian influenza): Latest situation in England: https://www.gov.uk/government/news/bird-flu-avian-influenza-latest-situation-in-england (Accessed 24/12/22).

63   Express: Bird flu symptoms: How many people died of bird flu? Is bird flu contagious?: https://www.express.co.uk/life-style/health/1368981/bird-flu-symptoms-how-many-people-died-bird-flu-is-bird-flu-contagious-evg (Accessed 24/12/22).

64   Hawk Conservancy Trust: https://www.hawk-conservancy.org/ (Accessed 26/07/23).

65   Taubenberger, J.K. and Morens, D.M. (2017) H5Nx panzootic bird flu – influenza's newest worldwide evolutionary tour. *Emerging Infectious Diseases, 23*, 340.

66   Science: Incredibly concerning: Bird flu outbreak at Spanish mink farm triggers pandemic fears: https://www.science.org/content/article/incredibly-concerning-bird-flu-outbreak-spanish-mink-farm-triggers-pandemic-fears#.Y9J8jwFJSSY.twitter (Accessed 31/01/23).

67   The Atlantic: We have a mink problem. Birds aren't humanity's only bird-flu worry: https://www.theatlantic.com/health/archive/2023/02/mink-farm-bird-flu-virus-infection-spread/673236/ (Accessed 12/04/23).

68   BBC: Bird flu 'spills over' to otters and foxes in UK: https://www.bbc.co.uk/news/science-environment-64474594 (Accessed 14/02/23).

69   SMC: H5N1 avian influenza: An old acquaintance that is changing fast: https://sciencemediacentre.es/en/h5n1-avian-influenza-old-acquaintance-changing-fast (Accessed 27/02/23).

70   Yahoo!News: Bird flu kills hundreds of sea lions in Peru: https://uk.news.yahoo.com/bird-flu-kills-hundreds-sea-213949358.html (Accessed 27/02/23).

71   The Daily Mail: Bird flu 'may mutate to kill more than 50% of humans': https://www.msn.com/en-gb/health/other/bird-flu-may-mutate-to-kill-more-than-50-of-humans/ar-AA17sI02 (Accessed 15/02/23).

72 Guardian: Bird flu: 11-year-old girl in Cambodia dies after being infected: https://www.theguardian.com/world/2023/feb/24/bird-flu-11-year-old-girl-in-cambodia-dies-after-being-infected (Accessed 27/02/23).

73 Nature: Girl who died of bird flu did not have widely-circulating variant: https://www.nature.com/articles/d41586-023-00585-1?utm_source=Nature+Briefing&utm_campaign=bce4cc3e93-briefing-dy-20230228&utm_medium=email&utm_term=0_c9dfd39373-bce4cc3e93-45565222 (Accessed 01/03/23).

74 Guardian: China records first H3N8 bird flu death, WHO says: https://www.theguardian.com/world/2023/apr/12/china-records-first-h3n8-bird-flu-death-who-says (Accessed 12/04/23).

75 Moore, M.J., Rathish, B. and Zahra, F. (2022) Mpox (Monkeypox). In *StatPearls* [Internet]. StatPearls Publishing, Treasure Island, FL.

76 CDC: About Mpox: https://www.cdc.gov/poxvirus/monkeypox/about/index.html (Accessed 27/02/23).

77 Newsweek: How contagious is monkeypox? R Number vs. COVID: https://www.newsweek.com/monkeypox-r-number-compared-covid-how-contagious-1709975 (Accessed 27/02/23).

78 UK Government: Mpox (monkeypox): Background information: https://www.gov.uk/guidance/monkeypox#clinical-features (Accessed 24/12/22).

79 The Conversation: How monkeypox epidemic is likely to play out – in four graphs: https://theconversation.com/how-monkeypox-epidemic-is-likely-to-play-out-in-four-graphs-184578 (Accessed 27/02/23).

80 Nature: WHO may soon end mpox emergency — but outbreaks rage in Africa: https://www.nature.com/articles/d41586-023-00391-9?utm_source=Nature+Briefing&utm_campaign=4cb84e684e-briefing-dy-20230214&utm_medium=email&utm_term=0_c9dfd39373-4cb84e684e-45565222 (Accessed 27/02/23).

81 WHO: Monkeypox: https://www.who.int/news-room/fact-sheets/detail/monkeypox#:~:text=Monkeypox%20is%20usually%20a%20self,been%20around%203%E2%80%936%25 (Accessed 27/02/23).

82 Infection Control Today: Mpox emergence: A review of the 2022–2023 outbreak: https://www.infectioncontroltoday.com/view/mpox-emergence-review-2022-2023-outbreak (Accessed 27/02/23).

83 CDC: Mpox in animals: https://www.cdc.gov/poxvirus/monkeypox/veterinarian/monkeypox-in-animals.html (Accessed 27/02/23).

84 edie: World's first plastic-free PPE equipment launched to combat coronavirus pandemic: https://www.edie.net/worlds-first-plastic-free-ppe-equipment-launched-to-combat-coronavirus-pandemic/ (Accessed 16/02/23).

85 De-la-Torre, G.E., Rakib, M.R.J., Pizarro-Ortega, C.I. and Dioses-Salinas, D.C. (2021) Occurrence of personal protective equipment (PPE) associated with the COVID-19 pandemic along the coast of Lima, Peru. *Science of The Total Environment, 774*, 145774.

86 Financial Times: Scramble to secure PPE cost UK taxpayer extra £10bn: https://www.ft.com/content/bd7ea64c-7395-4547-acf4-47c71eea46e7 (Accessed 16/02/23).

87 The World Wildlife Fund: https://www.wwf.org.uk/ (Accessed 22/12/22).

88 Waltner-Toews, D. (2017) Zoonoses, one health and complexity: Wicked problems and constructive conflict. *Philosophical Transactions of the Royal Society B: Biological Sciences, 372*, 20160171.

89 Cleaveland, S., Sharp, J., Abela-Ridder, B., Allan, K.J., Buza, J., Crump, J.A., Davis, A., Del Rio Vilas, V.J., De Glanville, W.A., Kazwala, R.R. and Kibona, T. (2017) One Health contributions towards more effective and equitable approaches to

health in low-and middle-income countries. *Philosophical Transactions of the Royal Society B: Biological Sciences, 372,* 20160168.

90 WHO: One Health High-Level Expert Panel (OHHLEP): https://www.who.int/groups/one-health-high-level-expert-panel/meetings-and-working-groups (Accessed 15/08/23).

91 Sivasundaram, S. (2020) The human, the animal and the prehistory of COVID-19. *Past & Present, 249,* 295–316.

92 Braverman, I. (Ed.). (2022) *More-than-One Health: Humans, Animals, and the Environment Post-COVID.* Taylor & Francis.

93 One Health Commission: What is one health? https://www.onehealthcommission.org/en/why_one_health/what_is_one_health/ (Accessed 10/07/23).

94 CDC: History: https://www.cdc.gov/onehealth/basics/history/index.html (Accessed 10/07/23).

95 OHHLEP: The one health definition and principles developed by OHHLEP: https://cdn.who.int/media/docs/default-source/one-health/ohhlep/one-health-definition-and-principles-translations.pdf?sfvrsn=d85839dd_3&download=true (Accessed 10/07/23).

96 WHO: Prevention of zoonotic spillover: https://www.who.int/publications/m/item/prevention-of-zoonotic-spillover (Accessed 10/07/23).

97 WHO: Bureau's text of the WHO convention, agreement or other international instrument on pandemic prevention, preparedness and response (WHO CA+): https://apps.who.int/gb/inb/pdf_files/inb5/A_INB5_6-en.pdf (Accessed 21/07/23).

98 A phrase we borrowed from a BBC comedy drama: BBC: Blackadder: https://www.bbc.com/historyofthebbc/anniversaries/june/blackadder/ (Accessed 10/07/23).

99 WHO: 8th meeting of the One Health High-Level Expert Panel (OHHLEP) 11 & 12 November 2022- Singapore: https://cdn.who.int/media/docs/default-source/one-health/ohhlep/22_12_19_8th-full-panel-ohhlep-meeting-draft-report.pdf?sfvrsn=c9f6e58f_3&download=true (Accessed 10/07/23).

100 Lamichhane, J.R. and Reay-Jones, F.P. (2021) Editorial: Impacts of COVID-19 on global plant health and crop protection and the resulting effect on global food security and safety. *Crop Protection, 139,* 105383.

101 New Pro Containers: COVID-19, disinfectant and plants: https://www.newpro-containers.com/blog/covid-19-disinfectant-and-plants/ (Accessed 16/02/23).

102 Nature Portfolio: How plants could produce a COVID-19 vaccine: https://www.nature.com/articles/d42473-020-00253-2 (Accessed 16/02/23).

103 Mukherjee, P.K., Efferth, T., Das, B., Kar, A., Ghosh, S., Singha, S., Debnath, P., Sharma, N., Bhardwaj, P. and Haldar, P.K. (2022) Role of medicinal plants in inhibiting SARS-CoV-2 and in the management of post-COVID-19 complications. *Phytomedicine,* 153930.

104 Nature: WHO abandons plans for crucial second phase of COVID-origins investigation: https://www.nature.com/articles/d41586-023-00283-y?utm_source =Nature+Briefing&utm_campaign=4cb84e684e-briefing-dy-20230214&utm _medium=email&utm_term=0_c9dfd39373-4cb84e684e-45565222 (Accessed 16/02/23).

105 Chan, A. and Ridley, M. (2021) *Viral: The Search for the Origin of Covid-19.* Fourth Estate, London.

106 Worldometer: COVID live: https://www.worldometers.info/coronavirus/ (Accessed 22/12/22).

107 Wellcome Sanger Institute: Respiratory virus and microbiome initiative: Parasites and microbes: https://www.sanger.ac.uk/group/respiratory-virus-and-microbiome-initiative/ (Accessed 12/04/23).

108   Guardian: Plans for UK 'genomics transformation' aim to act on lessons of Covid: https://www.theguardian.com/society/2023/may/16/plans-for-uk-genomics-transformation-aim-to-act-on-lessons-of-covid (Accessed 16/05/23).

109   Agnelli, S. and Capua, I. (2022) Pandemic or panzootic – A reflection on terminology for SARS-CoV-2 infection. *Emerging Infectious Diseases*, *28*, 2552–2555.

110   WHO: Pandemic influenza preparedness framework: Partnership contribution high-level implementation plan III 2024-2030 (who.int) (Accessed 21/07/23).

Final thoughts

The discussion in this text has been a way to bring information about the COVID-19 pandemic and its effects on animals together in one place. Hopefully it will encourage a better understanding of how animals were affected by what was considered a human disease. For the future, we should be working towards being able, in the face of any potential new pandemic, to use the tools at our disposal (including analyses such as those presented earlier) to predict how animals will be affected, and how we might protect them (as well as how that information can help protect us).

It is clear that prevention would be better than needing a cure, and equally clear that a model of One Health which takes on board the health and welfare of animals as well as humans being important for its own sake (not just as it prevents human disease) is desirable, both morally and in terms of effective prevention. Such a model of prevention needs to be fully integrated with promoting a healthy natural environment (both healthy ecosystems "in the wild" and in the "built" environment). A fundamental underpinning for this must be an appropriate respect for all animal life, a recognition of each animal's place in the global ecosystem, and the role that a high welfare mindset will have in preventing or mitigating the conditions in which the next pandemic disease may arise.

Looking to future pandemics, aside from the potentially catastrophic effects on individual humans and human communities, we need to be mindful of the effects on individual animals and animal communities. We spend a lot of effort conserving species, but in a situation of severe, human-life threatening pandemic, we may end up taking measures such as culling, which, in some circumstances, could eradicate vulnerable species that may already be struggling from the burden of the disease. Mitigations

DOI: 10.1201/9781003427254-8

for our own health can be highly deleterious for animals with no recip-rocal benefit, for example squalene extracted for vaccine production but no vaccines licensed for animal use, right through to discarded PPE entangling and killing an animal, as has been seen to happen.

It seems inconceivable that we will not at some point experience another pandemic (the cosy false sense of well-being we had "pre-COVID" has been shattered). What remains to be seen is whether our newly opened eyes will focus on animals as something other than a source or vector of zoonotic disease or a laboratory animal model in the search to understand or cure it. Or will we, finally, come to understand that animals are participants with us in this game of life or death (along with the environment) where our health and well-being, or lack of it, is recip-rocally assured? We ignore that dynamic in future at our peril, as, after all, the animals simply share with us humans a single rock circulating in space.

Index

Printed in the United States
by Baker & Taylor Publisher Services

Printed in the United States
by Baker & Taylor Publisher Services